Environmental Peacemaking

Environmental Peacemaking

Edited by Ken Conca & Geoffrey D. Dabelko

WOODROW WILSON CENTER PRESS
Washington, D.C.

THE JOHNS HOPKINS UNIVERSITY PRESS
Baltimore and London

EDITORIAL OFFICES

Woodrow Wilson Center Press
One Woodrow Wilson Plaza
1300 Pennsylvania Avenue, N.W.
Washington, D.C. 20004-3027
Telephone 202-691-4010
www.wilsoncenter.org

Order from

The Johns Hopkins University Press
P.O. Box 50370
Baltimore, Md. 21211
Telephone 1-800-537-5487
www.press.jhu.edu

2 4 6 8 9 7 5 3 1

Library of Congress Cataloging-in-Publication Data

Environmental peacemaking / Ken Conca & Geoffrey D. Dabelko, editors.
    p. cm.
Includes index.
    ISBN 0-8018-7192-1 (hardcover : alk. paper) — ISBN 0-8018-7193-X
(pbk. : alk. paper)
    1. Environmental policy—International cooperation 2. Security,
International I. Conca, Ken. II. Dabelko, Geoffrey D. III. Title.
GE170 .E576637   2003
363.7'0526—dc21   2002014049

ABOUT THE CENTER

The Center is the living memorial of the United States of America to the nation's twenty-eighth president, Woodrow Wilson. Congress established the Woodrow Wilson Center in 1968 as an international institute for advanced study, "symbolizing and strengthening the fruitful relationship between the world of learning and the world of public affairs." The Center opened in 1970 under its own board of trustees.

In all its activities the Woodrow Wilson Center is a nonprofit, nonpartisan organization, supported financially by annual appropriations from the Congress, and by the contributions of foundations, corporations, and individuals. Conclusions or opinions expressed in Center publications and programs are those of the authors and speakers and do not necessarily reflect the views of the Center staff, fellows, trustees, advisory groups, or any individuals or organizations that provide financial support to the Center.

Contents

Tables, Maps, and Figure

Tables

Maps

Figure

Acknowledgments

This volume is the result of a series of workshops on problems and prospects of environmental peacemaking. As with all truly collaborative ventures, the group of people to whom we are grateful extends far beyond the authors whose work is represented in these pages. Our thanks to the many individuals who attended the workshops, commented on the chapters, or otherwise contributed their time and thoughts along the road to the final product. Our sincere thanks in this regard to Matthew Auer, Anna Brettell, Ken Cousins, Nils Petter Gleditsch, Natalie Goldring, Scott Hajost, Allen Hammond, Peter Hayes, Thomas Jandl, Thomas Lovejoy, Allison Morrill, Tadashi Okimura, D. J. Peterson, Dennis Pirages, Marcus Schaper, Miranda Schreurs, Antoinette Sebastian, Indra de Soysa, Kate Watters, and Kenneth Wilkening.

Shanda Leather, Ariel Mendez, Jessica Powers, Jennifer Turner, and Clair Twigg were instrumental to the workshops held at the Woodrow Wilson Center. Allison Morrill, Jessica Powers, Joseph Brinley, Yamile Kahn, and Robert Lalasz helped greatly in the editing and assembly of the manuscript. Funding for these activities was provided by the Environmental Change and Security Project of the Woodrow Wilson International Center for Scholars and by the University of Maryland's Harrison Program on the Future Global Agenda. The Environmental Change and Security Project is especially thankful for the generous support of the U.S. Agency for International Development's Office of Population through a cooperative agreement with the University of Michigan Population Fellows Program, and of the Woodrow Wilson International Center for Scholars.

Finally, our most profound thanks are extended to the authors who contributed to this volume—for their enthusiastic participation, constructive criticism, and scholarly contributions.

About the Editors and Contributors

Ken Conca (editor) is an associate professor of government and politics at the University of Maryland, where he directs the Harrison Program on the Future Global Agenda. He is the author of *Manufacturing Insecurity: The Rise and Fall of Brazil's Military-Industrial Complex* (Lynne Rienner, 1997); editor, with Thomas Princen and Michael Maniates, of *Confronting Consumption* (MIT Press, 2002); editor, with Geoffrey D. Dabelko, of *Green Planet Blues: Environmental Politics from Stockholm to Kyoto* (Westview Press, 1998); editor, with Ronnie D. Lipschutz, of *The State and Social Power in Global Environmental Politics* (Columbia University Press, 1993); and author, with Wayne Sandholtz et al., of *The Highest Stakes: The Economic Foundations of the Next Security System* (Oxford University Press, 1992).

Geoffrey D. Dabelko (editor) is the director of the Environmental Change and Security Project at the Woodrow Wilson International Center for Scholars. He is editor of the annual *Environmental Change and Security Project Report* and editor, with Ken Conca, of *Green Planet Blues: Environmental Politics from Stockholm to Kyoto* (Westview Press, 1998).

Douglas W. Blum is a professor of political science at Providence College and adjunct professor of international studies at the Thomas J. Watson, Jr., Institute of International Studies at Brown University. Much of his recent work has focused on the Caspian region, and he has published and spoken on a number of related themes including Russian policy, American policy, energy geopolitics, and environmental security in the Caspian basin. He is currently working on a monograph titled "State, Society, and Cultural

Globalization: The Discourse of National Identity in the Transcaspian Region."

Pamela M. Doughman is an associate energy specialist at the California Energy Commission and a research associate of the Harrison Program on the Future Global Agenda at the University of Maryland. She is a contributing author to Joachim Blatter and Helen Ingram, eds., *Reflections on Water: New Approaches to Transboundary Conflict and Cooperation* (MIT Press, 2001).

Ashok Swain is an associate professor in the Department of Peace and Conflict Research at Uppsala University (Sweden) and director of Uppsala University's Programme of International Studies. He is the author of *The Environmental Trap: The Ganges River Diversion, Bangladeshi Migration, and Conflicts in India* (Uppsala University Department of Peace and Conflict Research, 1996) and, with Peter Wallensteen, of *International Fresh Water Resources: Conflict or Cooperation?* (Stockholm Environment Institute, 1997).

Larry A. Swatuk is a senior research fellow at the Centre for Southern African Studies of the University of the Western Cape, where he is coordinator of the research project titled "Security, Ecology, Community: Contesting the Water Wars Hypothesis in Southern Africa." Dr. Swatuk also lectures in the Department of Politics at the University of Botswana. His most recent publication is the co-edited collection (with Peter Vale and Bertil Oden) *Theory, Change, and Southern Africa's Future* (Palgrave, 2001).

Stacy D. VanDeveer is an assistant professor of political science at the University of New Hampshire. In addition to authoring and co-authoring a number of articles, working papers, and reports, he is editor, with L. Anathea Brooks, of *Saving the Seas: Values, Science, and International Governance* (Maryland Sea Grant Press, 1997) and editor, with Geoffrey D. Dabelko, of *Protecting Regional Seas: Developing Capacity and Fostering Environmental Cooperation in Europe,* conference proceedings (Washington, D.C.: Woodrow Wilson International Center for Scholars, 2000).

Erika Weinthal is an assistant professor of political science at Tel Aviv University. She is the author of *State Making and Environmental Protection: Linking Domestic and International Politics in Central Asia* (MIT Press, 2002).

1

The Case for Environmental Peacemaking

Ken Conca

Beyond Ecological (In)security

Fears that a crowded, resource-stressed planet could be the site of increasing conflict and violence are certainly not new; futurists such as Fairfield Osborn and Harrison Brown voiced such warnings in the early 1950s.[1] But only recently has this possibility become the subject of careful, critical evaluation. The past decade has seen the emergence of a large body of research examining possible links between environmental degradation and violent conflict,[2] as well as a parallel discussion in policy circles about the ecological sources of insecurity.

Most scholars remain skeptical of the idea that environmental change has been, or is soon to be, an important cause of war between nations. But several have argued that there is a dangerous and growing connection between environmental change and violent outcomes on a local or regional scale; these outcomes include episodes that can spill across borders. Environmental problems are most combustible when they exacerbate existing social tensions based on class, region, or ethnicity. When such tensions are triggered in the absence or weakness of social institutions that otherwise could mediate disputes or in the context of "failing" states, it is said, violent conflict may be triggered or worsened.

The possibility of links between environmental degradation and large-scale violence has spawned a highly polemical debate about "ecological security."[3] For some, ecological security means anticipating the violent results that might flow from pollution, resource scarcity, or ecosystem degradation and adapting traditional military and intelligence tools to counter these threats. According to John Deutsch, former director of the

1

U.S. Central Intelligence Agency, "national reconnaissance systems that track the movement of tanks through the desert, can, at the same time, track the movement of the desert itself. . . . Adding this environmental dimension to traditional political, economic and military analysis enhances our ability to alert policymakers to potential instability, conflict, or human disaster and to identify situations which may draw in American involvement."[4] For others, ecological security means a much more fundamental rethinking of what we mean by security in a tightly coupled world system; one implication of this view is the need to shift social resources away from traditional military means of defense and toward preventive economic and social measures.[5]

Despite its vagueness, or perhaps because of it, the concept of ecological security is beginning to take root in the policies and procedures of the industrialized world. The European Union (EU), the North Atlantic Treaty Organization, the Organization for Security and Cooperation in Europe, the Development Assistance Committee of the Organization for Economic Cooperation and Development (OECD), and the Japanese government have all recently commissioned state-of-the-art reports on the concept, with an eye toward developing policy guidelines and implementation procedures. The German government has proposed the creation of an "Ecological Security Council" within the United Nations (an idea raised earlier by former Soviet premier Mikhail Gorbachev). In the United States, government agencies including the State Department, the Defense Department, the Environmental Protection Agency, the Agency for International Development, and intelligence agencies have all begun to develop or apply ecological security mandates or policy guidelines. The EU has discussed ways of integrating the concept into its emerging common foreign and security policy, and, as the framework for the 2002 World Summit on Sustainable Development in Johannesburg was being conceived, the EU promoted ecological security as a summit theme.[6]

But the ecological security agenda has not played as well on the broader global stage. At best, it has been ineffective in engaging countries outside the OECD; at worst, it has been an obstacle to cooperation. The concept is widely seen by less developed countries as a rich-country agenda serving rich-country interests of access and control, and as a result it has made no serious headway in global forums. Other governments have long feared that increased interest in global environmental protection on the part of the United States and Europe might hamper their own quest for economic development.[7] Furthermore, the environmental priorities of less developed countries in international forums have been different from those of the North, the former stressing urban air and water quality, sanitation, and toxic contamination as opposed to global systems of climate, ozone, and the planet's stock of biodiversity. Recasting already contentious environmental debates in security terms—with the problematic connotations of intervention, unequal power, and a lack of voice in global institutions—has proven to be a diplomatic nonstarter. Consider a recent public statement by Kader Asmal, chair of the World Commission on Dams and previously South Africa's minister of water affairs and forestry:

> With all due respect to my friends, have battles ever been fought over water? Is water scarcity a casus belli? Does it divide nations? The answer is no, no and no. Indeed, water, by its nature, tends to induce even hostile co-riparian countries to co-operate, even as disputes rage over other issues. The weight of historical evidence demonstrates that organised political bodies have signed 3600 water-related treaties since AD 805. Of seven minor water-related skirmishes, all began over non-water issues.[8]

Asmal's statement, which reflects the skepticism often voiced by Southern elites, was made in response to a growing chorus of voices concerned that water shortages, water pollution, and other forms of environmental change are likely to trigger violent social conflict.[9]

We thus face a conundrum: Ecological security is emerging within OECD member nations as a powerful frame for international environmental protection, yet its terms of reference constitute an obstacle to international cooperation in the very places where the ecological insecurities of people and communities are most starkly displayed. In our judgment, the central reason for this is that neither environmental-conflict research nor ecological-security polemics have provided a clear strategy for *peace*. Studies warning of environmentally induced conflict typically end with highly generalized recommendations for environmental cooperation—but nearly always lack a careful analysis of the specific mechanisms or pathways by which such cooperation could be expected to forestall violence, mitigate conflict, or otherwise enhance the chances for peace, environmentally or otherwise. Divorced from any serious analysis of cooperative opportunities, the literature on environmentally induced conflict often reinforces a counterproductive, zero-sum logic of national security. It is as though the object were to anticipate where environmentally induced violence might occur, but not to figure out how to transform those situations through peaceful cooperation. It should be no surprise that this agenda has not engaged those countries suffering most immediately from ecological insecurities.

More transformative visions of ecological security, in contrast, often suffer from too much optimism rather than too little. To their credit, they seek to shift the focus from the security of states to the security of human beings in a world marked increasingly by threats without enemies.[10] But without research into specific pathways by which environmental triggers are to be defused or the zero-sum logic of resource conflicts is to be transformed, we are left with generalizations: that cooperation will automatically forestall conflict, or that "engaging" traditional security institutions on environmental matters will somehow produce a "greening" of security policies and practices.

Thus, for all that has been said and written, little has been done to investigate potentially important linkages between environmental conflict and environmental cooperation. It remains unclear whether and exactly how environmental cooperation can reduce the likelihood, scope, or severity of environmentally induced violence, or of violence and insecurity more generally. We have little knowledge of how to tailor environmental initiatives to speak specifically to the problem of violence. Even more important, we may be missing powerful peacemaking opportunities in the environmental domain that extend beyond the narrow realm of ecologically induced conflict. We know that international environmental cooperation can yield welfare gains, but can it also yield benefits in the form of reduced international tensions or a lesser likelihood of violent conflict? Such benefits could be a potentially powerful stimulus to environmental cooperation at a time when such a stimulus is badly needed.

The goals of this volume are to examine the proposition that environmental cooperation can generate synergies for peace, and to begin to identify the specific conditions and institutional forms through which those benefits might be realized. Although these questions present a challenging research puzzle—one we can only begin to unravel here—they are not of merely academic interest. We hope that a scholarly research agenda of environmental peacemaking can be an alternative to the current framing of "environmental security"—a frame that, in our view, cannot, in its current form, provide an adequate conceptual or normative basis for redressing the problems linking violence, nature, and human insecurity. We hope to challenge dominant constructions of environmental scarcity, ecologically induced violence, and state failure by stressing cooperative potential rather than violent inevitabilities, social transformation rather than social control, and peace rather than militarization. In essence, rather than seeking to pinpoint the environmental triggers of conflict, we are seeking to pinpoint the cooperative triggers of peace that shared environmental problems might make available.

Background: The Environment, Violence, and Insecurity

The ecological security debate has featured overlapping empirical and normative components. Empirically oriented work has focused primarily on environmental change as a potential cause of violent conflict. The research program led by Thomas Homer-Dixon at the University of Toronto has identified cases in which natural-resource depletion, ecosystem disruption, and other forms of environmental degradation appear to be linked to various types of intergroup conflict.[11] The Environment and Conflicts Project of the Swiss Federal Institute of Technology and the Swiss Peace Foundation has drawn a broadly similar picture.[12] These studies suggest that environmental change is best understood as a potential catalyst for conflict rather than as a sole or direct cause. There is very little evidence, for example, that environmental degradation has led directly to interstate war.[13] More support exists for the proposition that environmental problems can trigger or exacerbate localized conflicts along existing social cleavages such as ethnicity, class, region, or relative deprivation, and that these conflicts can in turn spill over to more widespread violence. In Homer-Dixon's words,

> Decreases in the quality and quantity of renewable resources, population growth, and unequal resource access act singly or in various combinations to increase the scarcity, for certain population groups, of cropland, water, forests, and fish. This can reduce economic productivity, both for the local groups experiencing the scarcity and for the larger regional and national economies. The affected people may migrate or be expelled to new lands. Migrating groups often trigger ethnic conflicts when they move to new areas, while decreases in wealth can cause deprivation conflicts such as insurgency and rural rebellion. In developing countries, the migrations and productivity losses may eventually weaken the state which in turn decreases central control over ethnic rivalries and increases opportunities for insurgents and elites challenging state authority.[14]

Employing a different conceptual framework that stresses uneven patterns of modernization, Günther Bæchler reaches a broadly similar conclusion:

> Transformation of landscape leads to violent conflicts and wars if and when it accentuates structural heterogeneity, which tends to discriminate chiefly against those rural producers who are the victims of bad resource allocations, unequal resource distribution, high dependence on natural capital, and bad state performance outside the federal district or national capital areas. The widening gap between the modern sector, which encompasses poverty clusters in sensitive environments, constitutes a fault line. In many places this has become a real front line of environmentally caused violent conflicts.[15]

Among scholars, controversies continue to surround claims about environmentally induced conflict. Skeptics point to the long and tenuous causal chain in most environment-conflict models, with a myriad of social, economic, and political factors lying between environmental change and violent outcomes. They also point to a tendency to ignore the social and institutional roots of conflict in favor of simple and mechanistic models of "scarcity." Also, case-study research has gravitated toward cases involving both severe environmental degradation and widespread violent conflict (in other words, researchers have tended to choose only cases that have outcomes consistent with the underlying hypothesis).[16] Finally, many of the most frequently cited examples—including Haiti, Central America, parts of South and Southeast Asia, and eastern Africa—involve places that seem woefully overdetermined for conflict, given the combination of social, ecological, political, and economic problems they face. This makes it hard to isolate the causal role of environmental drivers or the relative importance of environmental considerations.

Perhaps in response to these criticisms, a subsequent wave of case-study projects has begun to appear. A follow-up study titled "Environmental Scarcities, State Capacity, and Civil Violence" was undertaken at the University of Toronto, focusing on societal responses to ecological stress in China, India, and Indonesia. The findings stress the role of "environmental scarcities" in increasing financial and political demands on the state, provoking predatory behavior on the part of political and economic elites, mobilizing marginalized groups reliant on natural resources for their livelihoods, and undermining state revenues as economic productivity declines.[17] The Swiss Peace Foundation launched projects stressing the management of environmental conflicts in the Horn of Africa and the Nile River basin. These studies have emphasized the mapping of different stakeholder perceptions and seek to catalyze networking activities that facilitate the management of resulting resource conflicts.[18]

Given the controversies surrounding the case-study approach, there also have been efforts to test environment-conflict linkages more systematically, through cross-national quantitative studies. Wenche Hauge and Tanja Ellingsen found some support for theses of environmentally induced violence.[19] They found that environmental degradation has a positive effect on conflict, particularly at lower levels of violence. However, the effect was not as strong as that of key political and economic variables (specifically, the level of economic development and the type of governing regime). One way to interpret these results is that environmental change can be associated with violent outcomes, but that these effects are mediated by fundamental economic and political conditions in society. The State Failure Task Force, sponsored by the U.S. government, also produced a quantitative study with mixed findings. The task force found no direct relationship between environmental degradation and "state failure" (a term they defined to include certain extreme forms of violence, including revolutionary war, ethnic war, adverse or disruptive regime changes, and genocide).[20] The task force did uncover some potentially important indirect linkages, however. One of these was a significant link between environmental degradation and infant mortality, which the authors took as a proxy for broader quality-of-life conditions. Also, states suffering from high vulnerability and low state capacity were more likely to experience "failure" in the context of high deforestation rates than when such rates were low. This finding was interpreted to mean that the problem is not environmental degradation per se, but rather environmental degradation in the context of a low capacity to respond.

In principle, quantitative studies offer a useful way to test the merits of the common criticisms of environment-conflict research, because they make it possible to test very large numbers of cases and to control for other potential causes of conflict. In practice, the quantitative approach is plagued by problems of data quality and causal inference. Studies tend to be based on fragmentary and incomplete environmental data that may not be comparable across different societies. Researchers have little choice but to use national-aggregate data, which masks the internal inequalities and subnational processes that case-study research shows to be of central importance on the path to violence. Also, key social and political variables such as "state capacity" are difficult or impossible to capture with off-the-shelf quantitative indicators.

A more explicitly normative strand of the debate has focused on whether and how to incorporate "environmental threats" into security policies. This portion of the debate has been poorly focused and highly polarized, in large part because security is an essentially contested concept.[21] The objects of security may vary from individuals to social groups to political regimes to nation-states to global society. The threats toward which security policies are directed may be defined quite narrowly in terms of external military forces; they may also be conceived much more broadly to include risks to human health, standards of living, or community well-being. Traditionalists stress the incorporation of environmental considerations into the unreconstructed workings of the national security state, whereas more transformative approaches to ecological security stress reorienting societal resources, separating the idea of threat from the idea of enemy, and generally "greening" and demilitarizing security discourse, policies, and institutions.[22]

Both the traditionalist and the transformative frames of reference for ecological security have been justly criticized as ambiguous, vague, and contradictory.[23] A narrow security-state frame of reference ignores several inconvenient facts: the poor fit between military tools and the environmental "threats" they are meant to counter; the enormous ecological toll of war, war preparation, and national-security institutions; and the obstacles to effective environmental cooperation created by the zero-sum logic of the national-security state.[24] The more transformative vision of environmental security has also been subjected to intense criticism. Skeptics point to what they see as a naive belief that engaging security institutions on environmental matters will "green" security rather than militarize environmental policy. There is also a troubling tendency to wield the concept of security in a way that is at once both vague and all-encompassing. The use of the powerfully evocative but poorly specified concept of development, which has become a vehicle for so many contradictory agendas, offers a cautionary tale about this sort of terminological redefinition. And the specific pathways by which "ecological security" will not only reduce environmental triggers of conflict but also transform the zero-sum logic of the national security state too often remain unclear or implicit.

Conceptualizing Environmental Peacemaking

Our starting point is to stand the core premise of ecological (in)security on its head: Rather than asking whether environmental degradation can trigger broader forms of intergroup violent conflict, we ask whether environmental cooperation can trigger broader forms of peace. Peace can be thought of as a continuum ranging from the absence of violent conflict to the inconceivability of violent conflict. Moving along this continuum demands a set of social transformations related both to people's material livelihoods and to their identity-based associations. The material challenge of peace is to address problems of structural violence and social inequality—problems that make intergroup violence, on scales from local to transnational, all too easy to imagine.[25] In identity terms, peace can be equated with building an "imagined security community," in which new identity constructs emerge whose borders are defined not by territorial state boundaries but rather by shared norms of peaceful conflict resolution and "dependable expectations of peaceful exchange."[26]

Our purpose is to ask whether environmental cooperation can generate movement along the peace continuum, rendering violent conflict less likely or less imaginable. In other words, our interest in environmental peacemaking goes far beyond simply forestalling environmentally induced conflict, to ask whether environmental cooperation can be an effective general catalyst for reducing tensions, broadening cooperation, fostering demilitarization, and promoting peace. Admittedly, such claims are difficult to test. One could measure environmental cooperation through participation in international environmental agreements and look for a correlation between such participation and peaceful international behavior. Such a correlation might mean that environmental cooperation helps to build peace—but it could just as easily mean that peace is a prerequisite for environmental cooperation, not a result of it. Furthermore, it is generally recognized that governments are willing to enter into formal international agreements when the costs of doing so are low, that many environmental accords lack significant power to bind states, and that many forms of environmental cooperation are not codified in formal agreements. These facts make readily available indicators such as ratification of environmental treaties or participation in international environmental regimes poor measures of meaningful cooperation.[27]

Despite such difficulties, a strong theoretical basis supports the proposition that environmental collaboration can have positive spin-offs for peace.

The basis for this claim lies partly in the general conditions understood to facilitate cooperation, partly in the issue characteristics common to many environmental problems, and partly in the kinds of social relations that are engendered by ecological interdependencies. A recent review of the literature on cooperation theory by one of the editors of this volume suggests two general pathways by which environmental peacemaking might occur: changing the strategic climate and strengthening post-Westphalian governance.[28] The first path involves transforming the more immediate problems of mistrust, uncertainty, suspicion, divergent interests, and short time horizons that typically accompany conflictual situations. Along this pathway, the premise of environmental peacemaking would be to alter these dynamics by exploiting environmental problems as opportunities. Thus, the technical complexity of many environmental issues might be seized as an opportunity to create cooperative knowledge. Overlapping ecosystemic interdependencies might provide a chance to create opportunities for shared gains and establish a tradition of cooperation. Environmental challenges often demand anticipatory responses; the full scope of the effects of pollution and ecosystem degradation, for example, may not be apparent until natural systems are at or near the crisis point. Under these circumstances, it might be possible to push actors to lengthen "the shadow of the future"—that is, to extend the time horizon that frames the bargaining process. We make no claim that environmental problems are unique in having these opportunistic features—simply that they do present a rich set of opportunities.

A second or "post-Westphalian" pathway views peace not simply as stable interstate relations but, more broadly, as a shared collective identity within which violent conflict becomes inconceivable. Thus, the focus turns from the dynamics of interstate bargaining to the broader patterns of trans-societal relations. In this context, a strategy of environmental peacemaking would emphasize creating and exploiting positive forms of trans-societal interdependence, building transnational civil-society linkages, fostering new norms of environmental responsibility and peaceful dispute resolution, and transforming opaque, security-minded institutions of the state.

In other words, a robust strategy of environmental peacemaking must work on at least two different levels. First, it must create minimum levels of trust, transparency, and cooperative gain among governments that are strongly influenced by a zero-sum logic of national security. Second, it must lay the foundation for transforming the national-security state itself, which is too often marked by dysfunctional institutions and practices that become further obstacles to peaceful coexistence and cooperation. Both tasks are necessary. Stabilizing intergovernmental relationships without promoting institutional transformation runs the risk of merely reinforcing the zero-sum statist logic of national security. If the state is not an instrument of genuinely sustainable development, rooted in environmental and social justice, then improving and stabilizing interstate dynamics is unlikely to transform situations of ecological insecurity. Creating trust, reciprocity, transparency, cooperative knowledge, shared gains, and habits of cooperation among such entities might simply yield more efficient plundering of resources. On the other hand, seeking to transform institutions while ignoring the need for stable interstate relations risks being ineffective, irrelevant, and potentially even counterproductive, given the huge toll that interstate hostilities and noncooperation can have on the environment and social well-being.

With regard to the first pathway, that of improving the contractual environment of intergovernmental bargaining, we see two core questions. First, does institutionalized environmental cooperation forestall the catalytic or triggering processes that figure prominently in scenarios of ecoviolence—and if so, exactly how, when, and why does this occur? We presume that it is not enough just to cooperate; both the form and the content of that cooperation are critical. Second, and more important, are there positive side effects from such cooperation that can create positive synergies for peace, in the form of trust-building, consensual knowledge, longer time horizons, the identification of mutual gains, or a growing habit of cooperation? The substantial literature on environmental cooperation that has emerged in the past decade has underscored the political role of certain key issue characteristics across a broad range of environmental problems: uncertainty, knowledge intensity, extended time horizons, highly nonlinear patterns of change, and overlapping ecosystemic interdependencies.[29] These characteristics can be understood as obstacles to cooperation: they may create conflicts of interest or perception, as well as barriers to collective action even where interests and perceptions converge. But we also know that it is possible to alter the contractual environment of interstate bargaining in ways conducive to environmental cooperation; a large body of hard and soft international environmental law now exists.[30] Can that transformation be carried out in ways conducive to peace, as well?

With regard to the second or "post-Westphalian" pathway, we direct our attention to the ways that environmentally based cooperation might spill outside of formal interstate channels to create trans-societal linkages or alter intergroup understandings and identities. Here we see three important sets of research questions. One set of questions involves the deepening and broadening of trans-societal linkages. What transnational ties, if any, have emerged or could emerge between local civil-society mobilizations around resource issues at the domestic level? In what places and around what issues have they coalesced most strongly? What aspects of human insecurity do they target, and for whom? Following Sidney Tarrow's conceptualization of transnational advocacy, are these linkages sufficiently continuous transnationally and sufficiently grounded domestically to constitute a transnational social movement?[31] We stress these questions in order to identify emerging or potential forms of transnational interdependence that create a stake in peaceful cooperation, beyond narrowly economic conceptions of regional integration.

A second cluster of questions focuses on the emergence or strengthening of regionally grounded identities and their relationships to particular forms of interdependence. Do environmental concerns mobilize constituencies that favor or oppose regional conceptions of community? What forms of institutionalized environmental cooperation reinforce what types of regional political and economic interdependence? Do shared regional identities find their basis in particular issues, such as water access or land use, or in more cross-cutting or holistic constructs? What role do formal institution-building and more diffuse processes of social mobilization play in this process? We stress these questions in order to identify types and forms of environmental cooperation likely to deepen the development of an "imagined security community" in which the violent resolution of conflicts becomes ever harder for actors to entertain.

A third cluster of questions addresses the transformation of state institutions in directions conducive to cooperative interaction, democratic accountability, and peaceful dispute resolution. How are state institutions responding to the emergence of something akin to a regional civil society? Do emerging practices of institutionalized cooperation, both formal and informal, challenge or reinforce the character of state organs with regard to societal participation, transparency, and social control? Does the state-society dynamic differ in its domestic and transnational expressions? The cases presented in this volume reveal complex links between, on the one hand, enhanced state capacity for environmental protection and, on the other, broader societal engagement in the process of environmental cooperation. Building state capacity and deepening the state's transnational linkages may transform state institutions and broaden stakeholder participation, but it cannot be assumed that these changes will be the result.

Regional Dimensions of Environmental Peacemaking

In raising these questions, we focus in particular on *regional* peace and cooperation, for both ecological and political reasons. It should be stressed that we do not see the core object of environmental peacemaking as removing the ecological causes of violent conflict, although that may be one important benefit. Indeed, our emphasis on regional politics, the problematic character of the state, and the need to transform interstate dynamics places us at a different level of analysis from the eco-violence literature, which tends to stress highly localized, subnational conflicts over environmental degradation as the trigger of violence. We are not suggesting that creating a regional dynamic of environmental peacemaking will forestall these localized challenges. Rather, we are asking whether and how environmental cooperation might be a useful instrument of international peace by removing multiple sources of insecurity, most of which are political, economic, and social rather than narrowly ecological.

We focus on the regional scale of international relations for several reasons. Relations of ecological interdependence tend to be more tangible, immediate, and apparent at the regional scale than when viewed globally. These characteristics suggest that it should be easier to see or envision the political effects of environmental cooperation in regional relations. Perhaps more importantly, the regional scale often affords a rich, tangible array of offsetting upstream/downstream and shared-commons relationships of ecological interdependence on which to build peace.

An important political rationale also points toward the regional focus. Although the idea of environmental peacemaking is not new, it has typically been applied to a few particularly difficult bilateral cases, such as water issues in the conflict between Israelis and Palestinians.[32] Shifting the focus from a bilateral to a regional scale makes sense because few governments think about security issues in purely bilateral terms.[33] More importantly, a regional focus takes account of the fact that regions as diverse as post-apartheid southern Africa, post-Soviet central Europe, and post–Cold War South Asia are currently in the process of sorting out new security relationships in the wake of a particularly turbulent period of change in the international system. By selecting cases in which states and societies are not locked into patterns of confrontation and conflict, but rather are groping toward new relationships under conditions of great uncertainty, the catalytic potential of environmental cooperation may be easier to envision.

To begin an examination of the problems and possibilities of regional environmental peacemaking, we present in this volume six case studies of the politics of regional environmental cooperation. Our cases involve an array of regional environmental issues: shared seas in the Caspian and Baltic seas; river politics in South Asia and in formerly Soviet Central Asia; a linked set of land, energy, and water issues in southern Africa; and pollution control along the U.S.-Mexico border. World regions differ greatly, of course, and so do the political dynamics surrounding different kinds of environmental issues. The structure of incentives shaping conflict and cooperation around shared river basins, for example, may differ dramatically from that of shared regional seas or land-based environmental issues. Our goal is to develop more sophisticated propositions about the problems and possibilities of environmental peacemaking; we hope the richness of these cases will allow us to move beyond general propositions by delving into the complexities of specific, real places.

One criterion for case selection was the presence of at least a nascent process of environmental cooperation to be studied in each region, allowing us to look for peacemaking possibilities. Several of the regions examined in this volume figure prominently in some of the more lurid popular depictions of a world beset by ecological insecurity and environmentally induced violent conflict.[34] We note that institutional arrangements for cooperative resource management have also begun to spring up in each region. Some are based in state policy, but many are also socially grounded; some are primarily domestic in focus, whereas others are inherently transnational in scope. Viewed collectively, these initiatives constitute important social experiments, on many different scales, in the collective management and shared governance of natural resources and ecosystems.

Here we stress an important caution: These cases should not be viewed as formal tests of hypotheses concerning environmental peacemaking, for the simple reason that governments in these regions and elsewhere have not given us enough cooperative activity to conduct such tests. The preliminary, halting, and tenuous character of environmental cooperation in these and other regions is such that we cannot conduct such tests rigorously. Instead, we are looking to amass evidence, however partial or indirect, that more aggressive environmental cooperation could create exploitable peacemaking opportunities. In other words, we hope to nudge governments, intergovernmental organizations, social movements, and other actors who have not been aggressive about environmental cooperation to be somewhat more so, by pointing to credible possibilities of peacemaking spin-offs, if they are willing to act to realize them.

In addition to choosing cases that display some interesting, if still nascent, developments, we also have chosen to focus on regions that have a sufficiently fluid security order to make it at least possible to imagine environmental cooperation acting as a catalyst for peace. None of the regions examined in this book is entirely peaceful. Yet each can be seen as groping toward a new set of relationships in the wake of recent change. The character of those changes may vary substantially: the end of apartheid in southern Africa, the collapse of the Soviet Union, the establishment of the North American Free Trade Agreement, or the impact of the end of the Cold War in South Asia. The common denominator is that of opportunities in the relatively fluid settings that have resulted.

A third selection criterion has been an effort to make sure that we are not merely looking for cases that seem to confirm our premise about peacemaking potential. We have chosen cases with wide differences in the prevalence, depth, and nature of conflict. This variation will help us to understand not only the possibilities but also the limitations of environmental peacemaking strategies. In particular, it helps us to identify other social, economic, or political factors that may be critical in shaping the link between environmental cooperation and peaceful change.

The resulting set of cases, while allowing us to gain insights and refine several of our premises, is obviously neither geographically nor ecologically comprehensive. We can easily imagine applying the same set of questions to biodiversity initiatives of the sort taking root in post-revolutionary Central America; to informal but emerging aspects of cooperation on air pollution in post–Cold War Northeast Asia; to transnational watershed relations in river basins as diverse as the Mekong, the Nile, and the Danube; or to the proliferation around the world of transnational "peace park" initiatives for regional-scale biological conservation.

Finally, we should note that we have been flexible in our understanding of what constitutes a "region." Clearly, shared resource systems and ecological interdependencies—airsheds, watersheds, border-crossing rivers, regional seas—can be seen at the heart of each of our cases. But the conception of a region in each case also embodies a political process, in the form of at least incipient institution-building for cooperation around those shared biophysical realities. We are asking in each case whether there is a plausible convergence of the ecological, political, and social underpinnings of regional cooperation—making it impossible to define the region in static terms or across any one of these dimensions.

Is There a Basis for Optimism?

Obviously, environmental cooperation does not occur easily or automatically. Regional problems can be particularly challenging: Of the world's 261 internationally shared rivers, fewer than 1 in 5 is the subject of a substantial international agreement on issues of environmental protection, shared management, or water allocation.[35] Nor will environmental cooperation automatically enhance peace in the ways posited above. Much depends on the specific institutional form of cooperation and the purposes it serves. But these cautions should not obscure the fact that many environmental problems have useful properties that can be seized upon to build peace while lessening ecological insecurities. In the past, skeptics have dismissed such contentions as mere "functionalism"—the type of naive thinking that was popular in the early days of the United Nations, before the harsh realities of the Cold War exposed the limits to peace through functional cooperation.[36]

We reject such out-of-hand dismissal on two counts. First, it is largely irrelevant that countries coordinating air travel or postal services do not automatically descend a slippery slope of ever-deepening cooperation, because these narrowly functional activities have little in common with environmental cooperation. Far from being the "low politics" of technical coordination, problems surrounding shared river basins, regional biodiversity, forest ecosystems, or patterns of land and water use are controversial, high-stakes questions that can and do engage the state at the highest levels. The case studies presented in this volume show quite clearly that the stakes in environmental cooperation typically are understood by all important actors to be quite high. This political salience may be a barrier or a spur to cooperation—but it is hardly akin to a narrowly functionalist approach in which trivial aspects of cooperation are somehow expected to translate into a robust peace.

Second, skeptics overlook the rise of the "new regionalism" as a byproduct of economic globalization and the highly fluid nature of many regional security orders in the post–Cold War world.[37] As suggested above, all of the regions examined in this volume are sorting out new security relationships in the wake of a particularly turbulent period of international change. None of the regions examined in this volume has experienced a democratically consolidated "end of history." Yet in each region, the political transformations of the 1990s—the end of the Cold War, the collapse of the Soviet Union, the demise of apartheid—have created space to at least contest dominant forms of resource extraction and systems of access to nature.

More generally, the regional effects of globalization are complex and by no means entirely healthy for ecological sustainability. But one important and potentially very healthy effect seems to be the moving of regional political dynamics out of narrow interstate forums and into a broader trans-societal context.[38] This new social space is an important site where we envision much of the process of environmental peacemaking occurring. It is well worth asking whether these changes create opportunities to build peace, lessen ecological insecurity, and break out of the zero-sum logic that so often plagues interstate relations.

Notes

1. Fairfield Osborn, *Our Plundered Planet* (Boston: Little, Brown, 1953); and Harrison Brown, *The Challenge of Man's Future* (New York: Viking, 1954).

2. Thomas F. Homer-Dixon, "On the Threshold: Environmental Changes as Causes of Acute Conflict," *International Security* 16, no. 2 (Fall 1991): 76–116; Thomas F. Homer-Dixon, "Environmental Scarcities and Violent Conflict: Evidence from Cases," *International Security* 19, no. 1 (Summer 1994): 5–40; Thomas F. Homer-Dixon, *Environment, Scarcity, and Violence* (Princeton: Princeton University Press, 1999); Reidulf K. Molvær, "Environmentally-Induced Conflicts? A Discussion Based on Studies from the Horn of Africa," *Bulletin of Peace Proposals* 22 (1991): 175–88; Günther Bæchler and Kurt R. Spillmann, eds., *Environmental Degradation as a Cause of War: Regional and Country Studies of Research Fellows* (Zurich: Rüegger, 1996); Günther Bæchler, *Violence through Environmental Discrimination* (Dordrecht: Kluwer, 1999); Paul F. Diehl and Nils Petter Gleditsch, eds., *Environmental Conflict* (Boulder: Westview, 2000); and Nils Petter Gleditsch, "Environmental Conflict and the Democratic Peace," in Gleditsch, ed., *Conflict and the Environment* (Dordrecht: Kluwer, 1997).

3. Jessica Tuchman Mathews, "Redefining Security," *Foreign Affairs* 67, no. 2 (Spring 1989):162–77; Daniel H. Deudney, "The Case against Linking Environmental Degradation and National Security," *Millennium: Journal of International Studies* 19 (Winter 1990): 461–76; J. Ann Tickner, *Gender in International Relations: Feminist Perspectives on Achieving Global Security* (New York: Columbia University Press, 1993); Ken Conca, "In the Name of Sustainability: Peace Studies and Environmental Discourse," *Peace and Change* 19, no. 2 (April 1994): 91–113; Ken Conca, "Peace,

Justice, and Sustainability," in Dennis C. Pirages, ed., *Building Sustainable Societies* (Armonk: M. E. Sharpe, 1996); Jyrki Käkonen, ed., *Green Security or Militarized Environment?* (Aldershot, U.K.: Dartmouth Publishing, 1994); Marc A. Levy, "Is the Environment a Security Issue?" *International Security* 20, no. 2 (Fall 1995): 35–62; Geoffrey D. Dabelko and David D. Dabelko, "Environmental Security: Issues of Conflict and Redefinition," *Environmental Change and Security Project Report* 1 (Spring 1995): 3–13; Ashok Swain, "Environmental Security: Cleaning the Concept," *Peace and Security* (Vienna) 29 (December 1997): 31–38; Alexander Carius and Kurt M. Lietzmann, eds., *Environmental Change and Security: A European Perspective* (Berlin: Springer, 1999); and Daniel H. Deudney and Richard A. Matthew, eds., *Contested Grounds: Security and Conflict in the New Environmental Politics* (Albany: State University of New York Press, 1999).

    4. John Deutsch, speech to the Los Angeles World Affairs Council, Los Angeles, July 25, 1996.

    5. For a review of this literature, see Dabelko and Dabelko, "Environmental Security: Issues of Conflict and Redefinition."

    6. See, for example, Alexander Carius et al., *The Use of Global Monitoring in Support of Environment and Security: Draft Final Report for the Directorate General Joint Research Centre of the European Commission* (Berlin: Ecologic, June 2000); European Union, "EU Non-Paper on the Preparation for the Review of Agenda 21 and the Programme for the Further Implementation of Agenda 21" (Brussels: European Union, 2000); and World Conservation Union (IUCN), "State-of-the-Art Review on Environment, Security, and Development Co-operation," draft report for the Working Party on Environment, Development Assistance Committee, OECD (undated mimeograph). See also the various editions of the *Environmental Change and Security Project Report*.

    7. Marian A.L. Miller, *The Third World in Global Environmental Politics* (Boulder: Lynne Rienner, 1995); Somaya Saad, "For Whose Benefit? Redefining Security," in Ken Conca and Geoffrey D. Dabelko, eds., *Green Planet Blues: Environmental Politics from Stockholm to Kyoto* (Boulder: Westview, 1998); Adil Najam, "An Environmental Negotiation Strategy for the South," *International Environmental Affairs* 7, no. 3 (Summer 1995): 249–87; and Tanvi Nagpal and Camilla Foltz, eds., *Choosing Our Future: Visions of a Sustainable World* (Washington, D.C.: World Resources Institute, 1995).

    8. Kader Asmal, quoted in "World Commission on Dams Chair Challenges 'Water War' Rhetoric," press release, World Commission on Dams, August 14, 2000.

    9. See Sandra L. Postel and Aaron T. Wolf, "Dehydrating Conflict," *Foreign Policy* no. 126 (September/October 2001): 60–67; and Peter H. Gleick, *The World's Water: The Biennial Report on Freshwater Resources 1998–1999* (Washington, D.C.: Island Press, 1998).

    10. Mathews, "Redefining Security"; Norman Myers, *Ultimate Security: The Environmental Basis of Political Stability* (New York: Norton, 1993); UN Development Programme, "New Dimensions of Human Security," *Human Development Report 1994* (New York: Oxford University Press, 1994); and Gwyn Prins, *Threats without Enemies: Facing Environmental Insecurity* (London: Earthscan, 1993).

    11. Homer-Dixon, "Environmental Scarcities and Violent Conflict: Evidence from Cases"; and Homer-Dixon, *Environment, Scarcity, and Violence.*

    12. The results of this project are summarized in a three-volume report: Günther Bæchler et al., *Kriegsursache Umweltzertörung: Ökologische Konflikte in der Dritten*

Welt und Wege ihrer friedlichen Bearbeitung (Zurich: Rüegger, 1996); Bæchler and Spillmann, eds., *Environmental Degradation as a Cause of War: Regional and Country Studies of Research Fellows;* and Günther Bæchler and Kurt R. Spillmann, eds., *Environmental Degradation as a Cause of War: Regional and Country Studies of External Experts* (Zurich: Rüegger, 1996). See also Günther Bæchler, "Why Environmental Transformation Causes Violence: A Synthesis," *Environmental Change and Security Project Report* 4 (Spring 1998): 24–44; and Bæchler, *Violence through Environmental Discrimination.*

13. This conclusion is not surprising when one considers the inherent difficulties in isolating the separate strands of cause and effect leading up to social events as complex as war. The conventional wisdom is that governments rarely perceive environmental issues as worth fighting over. But an alternate explanation for the lack of identifiable "environmental wars" is that wars are multi-causal events in which it is the convergence of circumstances, as opposed to any single identifiable factor, that results in violent conflict. This also helps to explain the inability of international relations scholars to develop an adequate general theory of war causation, despite the fact that the quest for such a theory has historically been the central preoccupation of the field.

14. Homer-Dixon, "Environmental Scarcities and Violent Conflict," 31–32.

15. Bæchler, *Violence through Environmental Discrimination,* xvi.

16. Nils Petter Gleditsch, "Armed Conflict and the Environment: A Critique of the Literature," *Journal of Peace Research* 35, no. 3 (May 1998): 381–400; and Levy, "Is the Environment a Security Issue?"

17. For a summary of the Project on Environmental Scarcities, State Capacity, and Civil Violence, see the website of the Peace and Conflict Studies Program at the University of Toronto: ⟨www.library.utoronto.ca/pcs/state.htm⟩.

18. For a description of the ECOMAN (focused on the Horn of Africa) and ECO-NILE (focused on the Nile basin) projects, see the project website maintained by the Swiss Peace Foundation at ⟨www.cx.unibe.ch/swisspeace/htm/eng_txt/re_en_e.html⟩. See also Colin H. Kahl, "Population Growth, Environmental Degradation, and State-Sponsored Violence: The Case of Kenya, 1991–93," *International Security* 23, no. 2 (Fall 1998): 80–119.

19. Wenche Hauge and Tanja Ellingsen, "Causal Pathways to Conflict," in Diehl and Gleditsch, eds., *Environmental Conflict,* 36–57.

20. The study tested the explanatory power of 75 different social, political, ecological, and economic variables for the set of 113 state failures occurring between 1957 and 1996, as well as a randomly selected control group of nonfailures during the same time period. The goal was to develop a model that could accurately identify cases of both state failure and nonfailure. Indicators of environmental degradation tested included deforestation, soil degradation, change in agricultural land, freshwater access, sulfur dioxide emissions, the fraction of available freshwater withdrawn, and population density. The study found no direct relationship between environmental degradation and state failure. Among the indicators tested, the best model for predicting state failure involved three variables: infant mortality, openness to international trade, and the level of democracy. The model employing these three variables allowed correct classification for two-thirds of the episodes in the study. See Daniel C. Esty et al., *State Failure Task Force Report: Phase II Findings* (McLean, VA: Science Applications International Corporation, 1998).

21. Arnold Wolfers, *Discord and Collaboration: Essays on International Politics*

(Baltimore: Johns Hopkins University Press, 1962). See also Ronnie D. Lipschutz, ed., *On Security* (New York: Columbia University Press, 1995); and Jon Barnett, *The Meaning of Environmental Security: Ecological Politics and Policy in the New Security Era* (London: Zed, 2001).

22. A useful overview of the debate can be found in Dabelko and Dabelko, "Environmental Security: Issues of Conflict and Redefinition." This and subsequent reports of the Environmental Change and Security Project, containing several examples from each genre in the environmental security debate, are available at the project's website: ⟨ecsp.si.edu⟩.

23. See Matthias Finger, "The Military, the Nation State, and the Environment," *The Ecologist* 21, no. 5 (September/October 1991): 220–25; Dabelko and Dabelko, "Environmental Security: Issues of Conflict and Redefinition"; Deudney, "The Case against Linking Environmental Degradation and National Security"; Conca, "In the Name of Sustainability"; Ken Conca, "The Environment-Security Trap," *Dissent* 45, no. 2 (Spring 1998); Käkonen, ed., *Green Security;* and Barnett, *The Meaning of Environmental Security.*

24. A notable exception is the aforementioned work of the Environment and Conflicts Project of the Swiss Federal Institute of Technology and the Swiss Peace Foundation, which includes research on cooperative resource management and conflict resolution in the Nile River basin and the Horn of Africa.

25. Johann Galtung, "The Basic Needs Approach," in Katrin Lederer, ed., *Human Needs* (Cambridge, U.K.: Oelgeschlager, Gunn, and Hain, 1980); John W. Burton, ed., *Conflict: Human Needs Theory* (London: Macmillan, 1990); and Barnett, *The Meaning of Environmental Security.*

26. Emanuel Adler, "Imagined (Security) Communities: Cognitive Regions in International Relations," *Millennium: Journal of International Studies* 26, no. 2 (Summer 1997): 249–77; Emanuel Adler and Michael Barnett, eds., *Security Communities* (Cambridge, U.K.: Cambridge University Press, 1998); and Michael N. Nagler, "What Is Peace Culture?" in Ho-Won Jeong, ed., *The New Agenda for Peace Research* (Aldershot, U.K.: Ashgate, 1999). See also Karl W. Deutsch et al., *Political Community and the North Atlantic Area* (Princeton: Princeton University Press, 1957).

27. George W. Downs, David M. Rocke, and Peter N. Barsoom, "Is the Good News about Compliance Good News about Cooperation?" *International Organization* 50, no. 3 (Summer 1996): 379–406. See also Ken Conca, "International Regimes, State Sovereignty, and Environmental Transformation: The Case of National Parks and Protected Areas," paper presented at the annual meeting of the American Political Science Association, San Francisco, August 1996.

28. For a more detailed discussion, see Ken Conca, "Environmental Cooperation and International Peace," in Diehl and Gleditsch, eds., *Environmental Conflict.*

29. See, for example, Peter M. Haas, *Saving the Mediterranean: The Politics of International Environmental Cooperation* (New York: Columbia University Press, 1990); Peter M. Haas, Robert O. Keohane, and Marc A. Levy, eds., *Institutions for the Earth: Sources of Effective International Environmental Protection* (Cambridge, Mass.: MIT Press, 1993); Martin List and Volker Rittberger, "The Role of Intergovernmental Organizations in the Formation and Evolution of International Environmental Regimes," in Arild Underdal, ed., *The International Politics of Environmental Management* (Dordrecht: Kluwer, 1997); John W. Meyer et al., "The Structuring of a World Environmental Regime, 1870–1990," *International Organization* 51, no. 4 (Fall 1997):

623–51; Ronald B. Mitchell, *Intentional Oil Pollution at Sea: Environmental Policy and Treaty Compliance* (Cambridge, Mass.: MIT Press, 1994); Oran R. Young, ed., *Global Governance: Drawing Insights from the Environmental Experience* (Cambridge, Mass.: MIT Press, 1998); Karen Litfin, "Sovereignty in World Ecopolitics," *Mershon International Studies Review* 41, no. 2 (November 1997):167–204; Kal Raustiala, "States, NGOs, and International Environmental Institutions," *International Studies Quarterly* 41, no. 4 (December 1997), 719–40; and Robert O. Keohane and Elinor Ostrom, eds., *Local Commons and Global Interdependence: Heterogeneity and Cooperation in Two Domains* (London: Sage, 1995).

30. United Nations (UN) Environment Programme, "Multilateral Environmental Agreements: A Summary," background paper presented by the UN Secretariat at the first meeting of the Open-Ended Intergovernmental Group of Ministers or Their Representatives on International Environmental Governance, New York, April 18, 2001, UN Document no. UNEP/IGM/1/INF/1 (March 30, 2001).

31. Sidney Tarrow, "Transnational Contention," in Tarrow, *Power in Movement: Social Movements and Contentious Politics,* 2d ed. (Cambridge, U.K.: Cambridge University Press, 1998).

32. See, for example, Miriam R. Lowi, *Water and Power: The Politics of a Scarce Resource in the Jordan River Basin* (Cambridge, U.K.: Cambridge University Press, 1993); Miriam R. Lowi, "Bridging the Divide: Transboundary Resource Disputes and the Case of West Bank Water," *International Security* 18, no. 1 (Summer 1993): 113–38; and Steven L. Spiegel and David J. Pervin, eds., *Practical Peacemaking in the Middle East,* Vol. 2: *The Environment, Water, Refugees, and Economic Cooperation and Development* (New York: Garland, 1995).

33. On the relationship between bilateral and regional security as analytic lenses, see Barry Buzan, *People, States, and Fear: An Agenda for International Security Studies in the Post–Cold War Era* (Boulder: Lynne Rienner, 1991).

34. See, for example, Robert Kaplan, "The Coming Anarchy," *Atlantic Monthly,* February 1994.

35. Calculated from data in the Transboundary Freshwater Disputes Database of Oregon State University; see ⟨terra.geo.orst.edu/users/tfdd/⟩. The estimate for the number of internationally shared river basins is from Aaron T. Wolf et al., "International River Basins of the World," *International Journal of Water Resources Development* 15, no. 4 (December 1999): 387–427.

36. David Mitrany, *The Functionalist Theory of Politics* (New York: St. Martin's, 1976); David Mitrany, *A Working Peace System* (Chicago: Quadrangle, 1966); Ernst B. Haas, *Beyond the Nation-State: A Functional Theory of Politics* (Stanford: Stanford University Press, 1964); Lowi, *Water and Power;* and Daniel Deudney, "Global Environmental Rescue and the Emergence of World Domestic Politics," in Ronnie D. Lipschutz and Ken Conca, eds., *The State and Social Power in Global Environmental Politics* (New York: Columbia University Press, 1993).

37. Edward D. Mansfield and Helen V. Milner, "The New Wave of Regionalism," *International Organization* 53, no. 3 (Summer 1999): 589–627; Timothy M. Shaw and Sandra J. MacLean, "The Emergence of Regional Civil Society: Contributions to a New Human Security Agenda," in Jeong, ed., *The New Agenda for Peace Research;* and Björn Hettne and Andras Inotai, *The New Regionalism: Implications for Global Development and International Security* (Helsinki: World Institute for Development Economics Research, UN University, 1994).

38. Shaw and MacLean, "The Emergence of Regional Civil Society"; and J. Lewis Rasmussen, "Peacemaking in the Twenty-First Century: New Rules, New Roles, New Actors," in I. William Zartman and J. Lewis Rasmussen, eds., *Peacemaking in International Conflict: Methods and Techniques* (Washington, D.C.: United States Institute of Peace Press, 1997).

2

Environmental Cooperation and Regional Peace: Baltic Politics, Programs, and Prospects

Stacy D. VanDeveer

Environmental change and security are increasingly, but often implicitly, linked in the countries around the Baltic Sea.[1] Even before recent debates, interest in academic and policy circles centered on new Baltic regional security conditions and structures created by the end of Cold War understandings of European security.[2] The contemporary Baltic regional security context raises, yet leaves unanswered, many questions of identity, membership, and institutional configuration. What constitutes security in the Baltic region and what (or who) constitutes the region itself? Environmental issues are part of these discussions. Analysis of the links between environment and security issues around the Baltic enhances understanding of both sets of concerns and offers opportunities to explore the importance of the much-neglected regional level within the growing environmental security literature.

Few works have explicitly studied the environmental aspects of contemporary Baltic and European security contexts. Notable exceptions do exist, however. In the late 1980s, Arthur Westing and his associates enunciated an environmental approach to comprehensive security for the Baltic.[3] Explicitly environmental efforts, such as the activities of the Helsinki Commission (HELCOM), beginning in 1974, illustrate the sustained interest of policymakers in regional cooperation. HELCOM's environmental efforts were said to offer opportunities for confidence building among the divergent states around the Baltic Sea.[4] More recently, environmental issues as security issues have become increasingly important components of the larger regional construction of new Baltic security conceptions. Nordic policymakers increasingly incorporate explicit references to environmental threats into national and regional "soft" security doctrines. Many environmental concerns (e.g., radioactive contamination) enter security debates on the eastern side of the Baltic Sea or in the Barents Sea region.

The growing literature on Baltic regional security issues is characterized by five overlapping themes: political-military issues; border definition, control, and management; economic issues associated with interdependence and postcommunist transitions; collective-identity issues, both regional and national; and environmental issues. Processes of national identity construction and concerns about ethnic minorities, refugees, and migration illustrate important links among these themes.[5] This region reveals the extent to which international institutions and initiatives dominate debate and policy regarding environmental security and highlight the growing interconnectedness of allegedly issue-specific international institutions.[6]

This chapter reviews the principal challenges to peace and environmental quality in the Baltic region. The major challenges to relatively peaceful international and domestic relations lie at the nexus of potential conflicts over unsettled borders, ethnic rivalries, and state failure. More specifically, many disputes over borders between formerly Soviet republics remain unsettled, ethnic tensions remain relatively high, and the "transitions" away from state socialism continue to be characterized by institutional instabilities. From an environmental standpoint, major challenges span the spectrum—from highly localized degradation and pollution on the one hand, to region-wide air and water contamination on the other. Furthermore, the presence in the region of relatively large quantities of radioactive materials (both civilian and military) and chemical and biological contaminants presents potentially important local and regional risks. These challenges connect environmental problems in the area usually understood as "the Baltic" to those in the Barents and polar regions. This chapter reviews the extensive regional environmental cooperation on the issues of marine environmental protection, nuclear power plant safety, and, to a lesser extent, air pollution. In general, these cooperation efforts are led and sponsored by the Western or Nordic countries in the region in attempts to raise regional environmental standards and implementation.

Pan-Baltic cooperation remains a—perhaps *the*—main source of regionally institutionalized cooperation. Regional environmental cooperation was institutionalized prior to the 1989–92 collapse of the region's communist regimes, and it continues to outpace attempts to construct region-wide economic and security institutions. Although regional economic and security cooperation often remains problematic among the Baltic region's postcommunist states in particular, environmental cooperation has led and often improved general bilateral and multilateral relations among the region's transition states. The continuing challenges to deepening economic and security cooperation underscore the fact that peaceful relations were not, and are not, a foregone conclusion around the Baltic. Regional security and economic cooperation have been impeded, in recent years, by such factors as historical animosities, ethnic tensions, border disputes, ineffective state institutions, and gaping economic inequalities.

Environmental cooperation in the region contributes to "changes in the strategic climate" and "strengthening post-Westphalian governance."[7] Importantly, this environmental cooperation takes place within the context of multiple areas of organized interstate and intersocietal regional cooperation. In other words, environmental cooperation is not a self-contained sphere of cooperation. In particular, the strengthening of incentives for regional environmental cooperation around the Baltic has occurred within the context of the drive toward greater pan-European economic and political integration. Nordic, Polish, and Baltic-state elites, in particular, remain nearly unanimous in their support for numerous forms of regional international cooperation (environmental, security, economic, cultural, and so on). Environmental cooperation around the Baltic contributes to regional peace by socializing transition-state actors—and hence transition states—by encouraging within state practice the institutionalization of norms related to democracy and human rights. This process parallels Ken Conca's notion of "post-Westphalian governance," one of the two pathways discussed in Chapter 1 through which environmental cooperation might be a catalyst for peace. However, Baltic environmental cooperation underscores the analytical difficulties associated with separating the influences in Conca's two paths in a region where both interstate and trans-societal environmental cooperation are expanding rapidly.

This chapter also sounds an important cautionary note regarding environmental cooperation and regional peace around the Baltic (and the Barents). "Regional" integration and issue-specific cooperation are asymmetrical. In almost every respect, Russia and Belarus remain less integrated and less involved than other states and societies in regional organizations and cooperative regimes. Baltic environmental security cannot be achieved or approximated without greater efforts to incorporate the interests and concerns of those two countries.

Political Cleavages and Potential Conflicts

The current status of the Baltic region as a location of generally peaceful international and civil relations is a relatively recent phenomenon. Its states and nations have long histories of enmity prior to the tense political and military competition of the Cold War era. Less than twenty years ago, the East German, Polish, and Soviet states all used forceful means to suppress dissent, raising fears of larger civil and international conflict.

The principal axes of real and potential conflict in the region involve the nexus of ethnic group antagonisms, border and boundary disputes, and the danger of worsening economic and political instability in transition states. These issues are simultaneously domestic and transnational in the former Soviet transition states of the Baltic region. In the Nordic states and Germany, prosperity and stability remain the rule. As illustrated in Table 2.1, Baltic regional states include some of the world's wealthiest countries and encompass great regional economic inequalities. Still, no state in the Baltic region ranks among the poorest third of the world's countries.

Table 2.1

Human Development Statistics for the Baltic Littoral Countries

Country	HDI rank[1]	Life expectancy at birth (years) 1998	Adult literacy rate (%) 1998	GDP per capita (PPP, U.S.$)[2] 1998
Norway[3]	2	78.3	99.0	26,342
Sweden	6	78.7	99.0	20,659
Finland	11	77.0	99.0	20,847
Germany	14	77.3	99.0	22,169
Denmark	15	75.7	99.0	24,218
Poland	44	72.7	99.7	7,619
Estonia	46	69.0	99.0	7,682
Lithuania	52	70.2	99.5	6,436
Belarus	57	68.1	99.5	6,319
Russia	62	66.7	99.5	6,460
Latvia	63	68.7	99.8	5,728

Source: United Nations Development Programme (UNDP), *Human Development Report 2000* (New York: Oxford University Press, 2000).

[1]Rank, out of a total of 174 countries, on the Human Development Index (HDI), a yardstick developed by the UNDP to rank countries according to their level of "human development." It is based on life expectancy, adult literacy, gross domestic product (GDP) per capita, and school enrollment ratios.

[2]Purchasing power parity (PPP) is an adjustment to income data to account for local price differences among countries.

[3]Norway, although not a Baltic Sea littoral state by most definitions, is nonetheless an active participant in Baltic regional environmental and security affairs and a central actor in Barents Sea cooperation.

Concern about maintaining peaceful relations among ethnic groups in the newly independent Baltic states predates their independence from the former Soviet Union. Fear that tensions between the Russian minorities and national ethnic groups in Estonia, Latvia, and Lithuania might beget violent conflict (both civil and international) have been expressed by policymakers and political analysts alike, inside and outside the Baltic region.[8] Since independence in the early 1990s, policymakers in these states have faced complex challenges of simultaneous state- and nation-building while preserving sovereign independence and reforming Soviet-style economic and political institutions. Whether the expansion of the North Atlantic Treaty Organization (NATO) to include the Czech Republic, Hungary, and Poland has brought more or less security and stability to the region remains unclear. Certainly the continuing desire of Estonia, Lithuania, and Latvia to enter NATO engenders heated debate around the region. Citizenship laws, particularly those associated with language requirements, and minority rights also remain controversial issues, particularly in Latvia and Estonia. Russian policymakers and political elites often express interest in protecting Russian minorities from discrimination and abuse in the Baltic states and other former Soviet republics. At times, this interest has been expressed (or interpreted) as threatening, and some politicians on both sides of this issue have resorted to blatant demagoguery.

Environmental politics and conditions have potential connections to the maintenance of peaceful civil relations among ethnic groups in the now-independent Baltic states. In Estonia and Latvia, for example, many of the most polluted industrial areas are populated predominantly by ethnic Russians. These residents face both the potentially greater health effects of pollution in these dirtier regions and costlier economic impacts from stronger environmental policies, privatization, or the restructuring of inefficient facilities and industries. Such areas include the oil shale industry in heavily Russian northeastern Estonia. Radically downsizing or closing down this inefficient and environmentally destructive industry would throw thousands of ethnic Russians out of work, potentially adding to their sense of alienation and discrimination in Estonia. In an illustration of ethnic politics in the Baltic states, ethnic Estonians and Latvians often describe predominantly Russian areas as "dirtier" than other parts of the country. Although there appears to be a trend toward more inclusive citizenship laws and more comprehensive minority rights vis-à-vis the Russian minorities in the Baltic states, informal or nonstate discrimination against Russians remains common.

Civil freedoms for environmentalists and for those critical of the military and organized crime in Russia constitute another problematic linkage between environmental politics and national identity. In one well-publicized example, Russian environmental activist Aleksandr Nikitin was imprisoned and tried on charges of espionage and disclosure of state secrets because he participated in a Bellona Foundation study of radioactive pollution risks associated with the Russian navy's Northern Fleet. St. Petersburg courts dismissed the case against Nikitin in October 1998 and then acquitted him in December 1999 (five years after his initial arrest and raids on Bellona's Russian offices), a decision eventually upheld on appeal.[9] Although the Russian courts appear to have maintained some independence from government pressures, President Vladimir Putin's position on these issues is unclear: while head of the state security service in 1999, for example, Putin accused environmentalists of providing "convenient cover" for foreign spying.[10]

Border disputes among the Baltic states and Russia remain embroiled in the political rhetoric that is typical of historically difficult relationships and stymied by the link between boundary delineation and access to resources (e.g., oil extraction, or river and sea access). In and of themselves, these unsettled border disputes are unlikely to spark violence. However, some danger exists when these disputes connect with existing ethnic tensions, growing economic inequalities, the temptation of many domestic political figures to use uncompromising rhetoric, and officials' desire to control the flow of people and goods across borders. As a result, international pressure from Western governments and institutions frequently focuses attention on the need for compromise and resolution of unsettled boundaries.

Regional stability, security, and environmental quality in the Baltic and Barents regions are all closely linked to the fate of institutional reform and restructuring in Russia. Russian social and political instability—including the possible failure of democracy and the further collapse of the economy—may imperil regional peace and stability. This danger, coupled with continued Russian exclusion from many Western institutions and fears in the West and in many postcommunist states of a resurrection of Russia's "imperial impulse," makes the situation in Russia a major regional security concern. Societal instability or violent conflict might induce large numbers of people to flee across borders, into states ill-equipped to handle them. The dire economic situation and the potential for socio-political instability also increase existing concerns regarding the safety of Russian nuclear weapons, the control of weapons-grade materials, and the overall management of radioactive materials.[11]

Potential Russian instability, although an oft-cited concern, is not the only potentially destabilizing factor. State and economic crises in other postcommunist states pose potential dangers as well. Belarus's "transition" continues to yield little tangible economic or political reform. Recent economic crises in Asia and Latin America suggest that overconfidence in the economic stability of Poland and the Baltic states could prove to be dangerous; the Baltic states have experienced a number of banking crises, and domestic economic inequality is growing rapidly.

Important trends in regional cooperation include the deepening and expanding of regional environmental cooperation under the auspices of HELCOM, across multiple levels of governance, and across public and private sectors; growing but asymmetric regional economic and political interdependence; and a tendency in most cases for Russia and Belarus to remain less integrated into the region's multilateral cooperation efforts. Primary barriers to intergovernmental and trans-societal cooperation in the region include the ongoing border disputes, resource inequality, and the multidimensional problem of low levels of state capacity (including but not limited to the problems of corruption and "rent-seeking" public-sector behavior). Border disputes between the three Baltic states and Russia inhibit environmental and other types of cooperation.[12] In addition, because of ethnic rivalries and historical tensions, often-uncompromising political rhetoric has inhibited agreements on border delineation. Several areas of very severe local environmental degradation lie within these disputed boundaries (or in watercourses that pass through disputed areas). Environmental cleanup, mitigation, and international assistance are all inhibited by unsettled boundaries. For example, in Estonia, Latvia, and Lithuania, boundary disagreements inhibit cooperation and implementation of HELCOM environmental policy recommendations (although Latvian officials have recently increased their efforts to resolve remaining boundary and border disputes). Latvian-Lithuanian relations remain particularly difficult, involving conflicts over sea and land borders, oil and mineral rights, and the environmental safety of proposed oil extraction and transport facilities in Lithuania. Latvian environmental officials in 1996 could name no ongoing international environmental cooperation between Latvia and Lithuania outside of programs under HELCOM auspices.[13]

Environmental Challenges

The primary environmental challenges in the Baltic and Barents regions include locally severe pollution and environmental degradation (which sometimes transcend national boundaries); marine pollution, particularly coastal pollution and ecological degradation; chemical contamination; radioactive contamination and the risks thereof; and atmospheric deposition of numerous pollutants. International environmental problems in the region include cross-border pollutant flows, damages to regional commons (e.g., marine quality and fish stocks), and social, economic, or political spillover of local effects (primarily in the transition states).

Almost eighty million people live within the industrialized, urbanized societies of the Baltic Sea drainage basin. Well over twenty million live on or near the coast. Estonia, Latvia, and Lithuania all lie entirely within the basin, while at least 90 percent of the territories of Finland, Poland, and Sweden are also situated within the Baltic Sea ecosystem. (Map 2.1 shows the Baltic Sea watershed superimposed on the existing political boundaries of the region.) The two other littoral states, Germany and Russia, are not situated primarily within the Baltic Sea catchment area, but they contribute significantly to regional economic output, pollution loads, and resource use. In addition, small portions of the Baltic Sea drainage basin lie in the nonlittoral states of Belarus, the Czech Republic, Norway, Slovakia, and Ukraine.

As a result of economic activity in affluent Western societies and in postcommunist "transition" societies, heavy ship and boat traffic (both commercial and pleasure) traverses the Baltic Sea. The sea's position on the border between East and West made it a major theater of the Cold War, thereby increasing naval activity as well. By the late 1980s, thirty naval bases were located on the Baltic Sea coasts and more than 1,900 vessels were deployed in the sea.[14] The exclusive economic zones and exclusive fishing zones of the Baltic Sea littoral states extend into the sea, dividing up virtually the entire marine area and its resources.

Baltic Sea environmental pollution is primarily a post–World War II phenomenon. With the dramatic upturn in economic activity and industrial production and the increasing use of intensive agriculture and forestry throughout Eastern and Western Europe in the late 1950s and 1960s, the effects of pollution loading into the Baltic Sea and the stress on the marine ecosystem became apparent. By the late 1960s, scientific experts in the region were convinced that the Baltic Sea was being dangerously degraded and that international cooperation would be required to address the problem.

Map 2.1
The Watershed of the Baltic Sea Drainage Basin

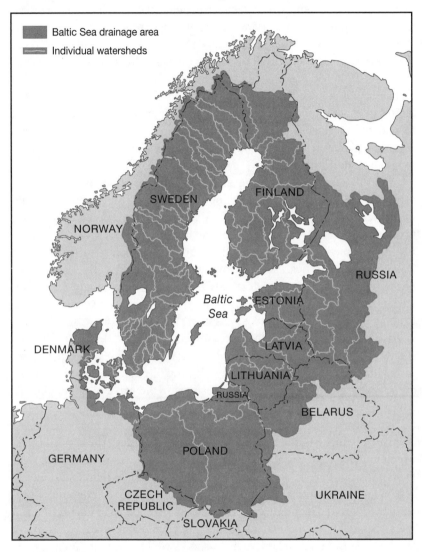

This regional scientific consensus was clarified and codified in a 1970 report sponsored by the International Council for the Exploration of the Sea.[15] As in the other parts of the industrialized world, the environmental damage was most apparent in localized areas of severe contamination.

The Baltic Sea has experienced a number of significant environmental changes, including increased levels of toxins, increased levels of nutrients (eutrophication), decreased oxygen levels, increasing salinity, and increasing temperatures.[16] Regional international cooperation has been most concerned with toxins, nutrients, and oxygen levels. These changes are largely (and most certainly) the result of anthropogenic influence. Other problems within the Baltic ecosystem that can adversely affect ecological quality and human health include the pollution of frontier rivers and boundary waters, radioactive pollution resulting from accidents and poor management of radioactive substances (most notably in the territory of the former Soviet Union), and atmospheric depositions such as acid rain and other forms of long-range air pollution.[17]

Currently, the most pressing environmental issue for the Baltic Sea ecosystem as a whole is the increasing eutrophication of marine waters and the associated decline in oxygen levels. Severe ecological degradation and toxic pollution exist in localized coastal areas. Certainly there exists room for improvement in marine-related environmental policy in the Western Baltic littoral countries. Nevertheless, in contrast to the early 1970s, most of the region's major multilateral environmental policy issues are associated with the environmental devastation left by state socialism and nutrient loading from agriculture and atmospheric deposition. How should such environmental "messes" in the transition countries be "cleaned up"? How should environmental policy and administration in these countries be reformed or created? Who should pay? Who will pay? These are the most prominent regional environmental questions in the wake of communism's collapse.

Nutrient levels, especially of nitrogen and phosphorus, have increased dramatically in the Baltic Sea over the last three decades. These substances, from municipal, industrial, and agricultural sources, enter the sea via river flows and atmospheric deposition. Point sources such as municipal sewage and industrial facilities are more easily controlled than nonpoint (agricultural) sources. Declining oxygen levels at lower depths of the Baltic Sea are associated with eutrophication; such oxygen declines adversely affect levels of micro-fauna and microorganisms as well as the reproduction rates of species such as cod.

Somewhat elevated levels of heavy metals, such as copper, lead, and cadmium, have been detected in sediments and living organisms in the Baltic Sea.[18] Dangerously high levels of these substances (and of mercury) are generally found only in localized coastal areas near industrial and municipal centers. The same is true of the pesticide DDT (dichlorodiphenyltrichloroethane), although its levels have been in decline since the mid-1970s. PCBs (polychlorinated biphenyls), which enter the Baltic Sea via river flows, ship paint, and atmospheric deposition, have stabilized at relatively low levels. Slow declines in PCB levels suggest that some littoral states may be violating restrictions placed on these compounds under the auspices of HELCOM.

In recent years, concern about pollutants from the dumping and spilling of chemical weapons and radioactive substances has risen greatly in the region. Generally, however, toxins have done the greatest environmental damage in coastal areas. This is especially true near industrial facilities such as pulp and paper mills, smelters, and fertilizer and other chemical producers, and around the sixty urban centers along the Baltic coast.

If eutrophication is a serious existing environmental challenge around the Baltic, radioactive contamination is likely the most serious potential or future challenge. In the neighboring Barents region, such chemical and radioactive contamination already exists at low levels, and there is a very real possibility that such pollution could get much worse. The Russian Northern Fleet's military bases and installations in this region have a history of dumping chemical and radioactive materials; large quantities of such materials remain inadequately contained and disposed of.[19] For the entire Baltic-Barents region, concerns about nuclear safety at power plants and weapons-production facilities, and the possibility of a repeat of the 1986 Chernobyl nuclear power plant accident (or worse), rank high on the priority lists of security planners and environmental policymakers.

In environmental security issues—as in many other areas—officials in Baltic states frequently look over their shoulders toward Russia.[20] The environmental legacy of Soviet central planning, including dramatic examples of terrestrial, hydrological, and atmospheric pollution, left fear and frustration among Baltic-state officials about the environmental dangers associated with being Russia's neighbors. During the period of negotiations and controversies surrounding Soviet troop withdrawal in 1992–94, Baltic-state officials and activists repeatedly raised the environmental impacts of ongoing Soviet military activities. The environmental legacies of Soviet-style communism in the Baltic states include poor wastewater treatment capacity; contaminated industrial cities; locally heavy coastal, riverine, lake, and groundwater pollution; and significant atmospheric deposition. The Soviet legacy also includes the memory of the Chernobyl disaster and its lingering radioactive contamination, as well as concern among policy-

makers and publics over environmental quality on and around former Soviet military bases. Such locations are associated with toxic and hazardous
wastes, groundwater pollution, and locally severe ecological degradation.
The environmental impacts of former Soviet bases resonate within Baltic
politics because the bases remain a symbol of decades of occupation.
Lastly, many public officials around the Baltic remain concerned about
potentially negative environmental and human health impacts from military
and industrial activities in Kaliningrad.[21]

Serious environmental problems—municipal and industrial wastes and
degradation resulting from oil shale extraction—exist in the areas around
Tallinn and the northeast region of Estonia. Pollution remains closely associated with Estonian understandings of Soviet exploitation and the Russians who came to work in the region. Lithuania's most significant environmental concerns are water pollution, nuclear safety at the Chernobyl-style
RBMK Ignalina nuclear power plant, and the cleanup of former Soviet
military installations. Despite domestic officials' prioritization of water
quality, nuclear safety at Ignalina has garnered much international and
domestic attention. Concern about the plant has been so strong that it has
diverted international attention away from other priority environmental
issues. This effect has been most pronounced in Sweden and Finland; it has
resonated less in Denmark.

Environmental Cooperation

Baltic regional environmental cooperation is extensive, including various
manifestations of formal, informal, intergovernmental, and nongovernmental collaboration. Such cooperation primarily involves and is driven by
actors from the region, although external actors have also been recruited.
Baltic environmental cooperation is highly institutionalized at multiple
levels of governance and has taken multiple forms: interstate (bilateral and
multilateral), inter-bureaucratic, subnational-to-subnational (regional and
local), scientific and technological, private sector, and nongovernmental
organization (NGO).

This chapter focuses mainly on state-driven forms of international environmental cooperation in the Baltic region. Nevertheless, regional cooperation, both multilateral and bilateral, exists among environmental
scientists, activist NGOs, and professional organizations (such as port authorities, cities, and other subnational governing bodies). Table 2.2, con-

Table 2.2

Observers of the Helsinki Commission

Governments	Belarus
	Ukraine
Intergovernmental organizations	
	Agreement on the Conservation of Small Cetaceans of the Baltic and North Seas (ASCOBANS)
	Intergovernmental Oceanographic Commission (IOC)
	International Atomic Energy Agency (IAEA)
	International Baltic Sea Fishery Commission (IBSFC)
	International Council for the Exploration of the Sea (ICES)
	International Maritime Organization (IMO)
	OSPAR Commission
	United Nations Economic Commission for Europe (ECE)
	United Nations Environment Programme (UNEP)
	World Health Organization, Regional Office for Europe (WHO/ EURO)
	World Meteorological Organization (WMO)
Nongovernmental organizations	
	Baltic and International Maritime Council (BIMCO)
	Baltic Ports Organization (BPO)
	BirdLife International
	Coalition Clean Baltic (CCB)
	European Chlor-Alkali Industry (EURO CHLOR)
	European Community Sea Ports Organization (ESPO)
	European Dredging Association (EuDA)
	European Fertilizer Manufacturers' Association (EFMA)
	European Union for Coastal Conservation (EUCC)
	International Council for Local Environmental Initiatives (ICLEI)
	International Network for Environment Management (INEM)
	Oil Industry International Exploration & Production Forum (E&P Forum)
	Standing Conference of Rectors, Presidents, and Vice-Chancellors of the European Universities (CRE)
	Stichting Greenpeace Council, Greenpeace International
	Union of the Baltic Cities (UBC)
	World Wide Fund for Nature (WWF)

taining the list of HELCOM observers, identifies many of these groups. Importantly, these groups function as transboundary conduits for specific types of expertise and for values, principles, and policy norms among the countries in the region.[22] Many of these groups engage in professional development or "capacity-building" activities with their members, promote public environmental education, and compile and distribute environmental information. Transnational norms are diffused into domestic spheres through discourse communities and social learning within regional networks and organizations. Furthermore, the networks established within such groups tend to overlap, often moving information, ideas, and norms across communities and organizations as well as within them.

Groups such as the Union of Baltic Cities, the ECOBALTIC Foundation, HELCOM's working group on public awareness and environmental education (a subsidiary of HELCOM's Programme Implementation Task Force), and the Baltic Sea Region On-Line Environmental Information Resources for Internet Access (BALLERINA) initiative all have public awareness and environmental education programs involving multinational NGOs and collaborations between NGOs and the public sector. These programs include narrowly tailored education and professional training programs as well as large and diffuse efforts to raise public awareness. For example, a number of states, universities, intergovernmental organizations, and NGOs are coordinating "An Agenda 21 for the Baltic Sea Region" to formulate and attempt to implement an action plan for sustainable development in the region.[23] As Table 2.2 suggests, Baltic environmental protection efforts constitute an institutionally dense web of connections—including parliamentarians and other policymakers, scientific and technical groups, advocacy NGOs of many kinds, and professional organizations from the region. In addition, many regional groups and organizations are linked to larger, pan-European and global institutions. Lastly, Norway occupies a kind of "half in, half out" status in Baltic regional organizations. Although Norway is not a member of HELCOM, many Norwegian organizations participate in and belong to other Baltic regional organizations and initiatives. Furthermore, Norwegian officials allocate some funds to Baltic protection efforts of various types.

The Helsinki Commission

For almost three decades, the Baltic Sea has been the focus of efforts to protect ecological quality and ensure the continued production of marine resources. Bilateral and multilateral environmental protection arrangements for the Baltic Sea date back to the late 1960s. In 1974 and again in 1992, representatives of the Baltic littoral states gathered in Helsinki to sign comprehensive environmental protection agreements.[24] The 1974 Helsinki Convention was the first regional international agreement limiting marine pollution from both land- and sea-based sources, whether air- or water-borne. The 1992 Helsinki Convention updated the 1974 agreement, expanding the treaty's scope and strengthening collaborative environmental policy. The Baltic Sea environmental protection regime was constructed and operated across the ideological and strategic divide between East and West; it became a model for other regional environmental protection regimes and conventions. Regional environmental cooperation around the Baltic Sea emerged from a poor strategic climate; divergent political, economic, and ideological systems, opposing military alliances, and East-West tensions made the Baltic region a tough case for international cooperation of any kind. It was initially blocked by Western states' refusal to even recognize the German Democratic Republic (East Germany). In negotiating the 1974 convention, some states in the region were so concerned with protecting their sovereignty and security that they exempted coverage of coastal waters—by far the most polluted—from the convention.

The 1974 Helsinki Convention established the Baltic Marine Environmental Protection Commission, or Helsinki Commission. HELCOM functions as a secretariat to administer and implement the convention. It began operation as an "interim commission" immediately after the signing. During the interim period (1974–80), Finland and Sweden provided the resources necessary to support the maintenance of international cooperation. Their early sponsorship and continued support have been essential for HELCOM's development. The convention came into force on May 3, 1980, following ratification by West Germany, the last of the seven signatories to ratify.[25]

HELCOM and its permanent committees and working groups operate on a one-country, one-vote principle. The chair of the commission rotates among all member countries for two-year terms. In theory, the costs of the commission are shared equally among the parties. In practice, Western states sponsor HELCOM programs and activities and Finland makes an extra yearly contribution to support the secretariat. HELCOM issues nonbinding environmental policy "recommendations" that require the unanimous support of the parties. State representatives with relevant scientific, technical, and legal expertise work out these recommendations in committee. Thus, HELCOM recommends common (regional) environmental policy standards and procedures to participant states.

HELCOM lacks formal enforcement powers and the convention makes implementation the responsibility of member states. Generally, states use national Baltic Sea committees or HELCOM offices to implement

HELCOM decisions.[26] The commission does not formally monitor compliance; rather, it coordinates environmental monitoring and national discharge reporting.

The early organization of HELCOM and its permanent committees, established under the auspices of the interim commission, was quite simple and small. For most of its history, HELCOM had three permanent committees (Maritime, Combating, and Scientific-Technological), as well as numerous working groups of experts. Subsidiary groups, usually committees, formulate recommendations and proposals for adoption by the full commission. The Maritime Committee advises the commission mainly on pollution from ships, offshore platforms, and waste disposal in ports. It works closely with the International Maritime Organization in London. The Combating Committee formulates rules and guidelines for combating spillage of oil and other harmful substances.

The Scientific-Technological Committee (STC) has dealt primarily with issues concerning the monitoring and assessment of pollution, its sources, and means of entry into the sea. The STC also promotes scientific and technical cooperation with relevant regional international bodies. The STC was itself subdivided into the Environment Committee, which concentrates on environmental monitoring and assessment, and the Technological Committee, which focuses on restricting point and nonpoint source pollutant discharges. Figure 2.1 illustrates HELCOM's structure throughout the 1990s. Ministers at the 1992 Diplomatic Conference adopted a resolution establishing a permanent Programme Implementation Task Force (PITF) to initiate, facilitate, and coordinate the implementation of the Baltic Sea program. As the permanent committee structure suggests, HELCOM's activities are of a highly scientific and technological nature. This characteristic allows HELCOM to take advantage of the authority and legitimacy gained by the perception that it existed apart from, or above, the normative environment of international politics and the ideological clash between East and West.

Increasing ministerial-level regional cooperation and collaboration has increased the "thick" institutional and organizational situation in Baltic regional cooperation. Peter Haas calls attention to regular ministerial conferences—every two or four years—as a "new institution" in Baltic regional environmental protection.[27] These conferences generally garner media attention and result in announcements of common goals, such as a call for 50 percent reductions in nutrient emissions into the Baltic Sea.

Figure 2.1

The Organizational Structure of HELCOM

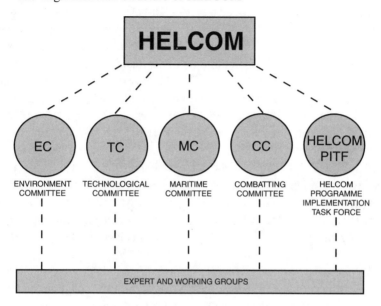

SEMINARS, SYMPOSIA, AD HOC GROUPS,
INFORMAL EXPERT MEETING

MEMBERS

DENMARK, ESTONIA, EUROPEAN COMMUNITY, FINLAND,
GERMANY, LATVIA, LITHUANIA, POLAND, RUSSIA, SWEDEN

However, environment ministers are not the only ones who now meet regularly around the region. In addition to regular meetings of prime ministers, ministers covering defense, health, transportation, economics/finance, and culture, to name a few, hold high-profile conferences. These ministers use these meetings to leverage bureaucratic power at home, setting interim and long-term goals for common policies.

HELCOM and its associated bodies have grown in size, scope, and specialization. HELCOM has adopted more than 175 recommendations, averaging more than 8 recommendations per year in the six years following the fall of communism (1992–98). This activity reflects the growing domestic and international interest in environmental protection in the Baltic region, growing transnational agreement on goals, and a greater willingness and ability on the part of the southern and eastern littoral states to engage in transnational environmental management.

The recommendations passed by HELCOM also are growing more stringent and precise. More than 45 HELCOM recommendations supersede or supplement earlier, less stringent, or less specific ones. All nine Baltic littoral states participate in regime activities, as do the Czech Republic, Slovakia, Norway, and the European Union (only the nine littoral states and the European Union are parties to the convention). Four international financial institutions (the European Bank for Reconstruction and Development [EBRD], the European Investment Bank [EIB], the Nordic Investment Bank [NIB], and the World Bank) are involved, as are the many observers listed in Table 2.2. In addition, the goals set at the 1988 and 1990 ministerial meetings and reiterated at the 1992 Diplomatic Conference demand more stringent environmental policies from HELCOM and its member states. Since 1988, most of HELCOM's activities have been aimed at implementing decisions made at ministerial conferences. The conferences, therefore, have altered regime dynamics.

The fact that the many regime changes—including the inception of periodic ministerial conferences, the signing of the 1992 convention, organizational expansion and restructuring, and the development of the Joint Comprehensive Environmental Action Programme—occurred simultaneously yielded mutual compatibility and a sort of synergistic momentum among these changes. HELCOM, the central organization of the regime and the bearer and promulgator of regime principles and norms, directed much of the regime change. HELCOM's role in the regime change is not predicted or well explained by mainstream international relations theories of regimes and organizations, which tend to treat international organizations as weak institutions that do not shape their associated regimes or state interests. Attention to regime principles and norms, however, does help to explain the nature of HELCOM's influence on regime change.

HELCOM has amassed a limited number of environmental and organizational successes. Oil inputs into the Baltic Sea have declined, as have concentrations of toxic substances such as DDT, PCBs, mercury, and cadmium in living organisms.[28] Populations of gray seals, ringed seals, harbor porpoises, and some bird species appear to be recovering slowly, though mostly on the northern side of the Baltic Sea.[29] International and transnational cooperation around HELCOM greatly increased the likelihood that vessels violating environmental regulations would be caught and held responsible.[30] It also increased coordination in combating accidents and minimizing their environmental damage. The exchange of technical information and knowledge became commonplace, intensifying steadily following the 1974 signing, and expanding greatly after the collapse of state-socialist governance. International cooperation has helped to reduce phosphorus and nitrogen loads, though these declines remain far short of those needed to reduce eutrophication and exceptional planktonic blooms.[31]

Despite a host of remaining environmental challenges, general agreement exists among Baltic officials and members of the regional scientific community that the Baltic Sea would have deteriorated into a more polluted body of water—with significant ecological and economic costs—in the absence of the 1974 Helsinki Convention and the ensuing forms of environmental cooperation. From its origins in Cold War détente, through the return of the Cold War's "chill" in the early 1980s and the collapse of Soviet-style communism, Baltic regional environmental cooperation emerged from an initially hostile strategic climate to become one of the most active, robust, and effective environmental regimes in the world. As recently as the early 1990s, this result and the persistence of peaceful international relations in the region remained in doubt. At that time, Soviet Moscow opposed the independence of the Baltic states, even as Baltic officials explicitly used HELCOM bodies and the 1992 treaty negotiations to assert their sovereign independence.

Toward Implementation

The Baltic Sea Joint Comprehensive Environmental Action Programme (JCP) identifies environmental problems and prioritizes action in all of the countries of the Baltic Sea catchment area. Its main focus, however, is on the most severely degraded areas in the former communist states. The JCP serves as a long-term outline for curative and preventive environmental action to clean up existing and ongoing ecological damage from point and nonpoint sources, promote sustainable development, and improve domestic legislation, regulation, institutional capacity, resource use, and financing for environmental activities.[32]

A high-level HELCOM task force compiled the JCP based on national plans drafted by all of the participant states, pre-feasibility studies, and special studies of specific ecological areas of concern, such as wetlands and agricultural drainage areas. International NGOs, including Greenpeace International and the World Wide Fund for Nature (WWF) commented on drafts of the pre-feasibility studies and the preliminary version of the JCP. Also commenting was Coalition Clean Baltic (CCB), a transnational umbrella group for local and national environmental NGOs from all of the Baltic littoral states. The CCB serves as a vehicle for environmental NGOs in Western Baltic states to support NGOs in transition states with modest levels of assistance. The CCB also functions as a forum for developing regional NGO consensus positions intended to influence domestic and international policymakers.

The JCP constitutes HELCOM's attempt to organize and support environmental activities in the postcommunist states and map out the required steps, as currently understood, for regionally comprehensive environmental protection of the Baltic Sea. The JCP covers six component areas of action: (1) policy, legal, and regulatory reform; (2) institutional strengthening and human resources development; (3) infrastructure investment; (4) management of coastal lagoons and wetlands; (5) applied research; and (6) public awareness and environmental education.[33] The total cost of the twenty-year program is estimated at 18 billion European currency units, or ECU (approximately $25 billion): 5 billion ECU for Phase I (1993–97), and 13 billion ECU for Phase II (1998–2012).[34] Phase I focuses on "priority hot spots," including detailed environmental assessments and audits and the beginning of mitigating, abatement, and elimination of these worst pollutant sources.

HELCOM does not raise the funding needed to implement the JCP. Rather, HELCOM intended the JCP to identify areas of need and legitimize them through expert scientific and technical assessment. HELCOM organizes resource-mobilization workshops, bringing together officials and private-sector actors in the postcommunist states with representatives of bilateral and multilateral donor organizations and prospective foreign investors.

The JCP has identified 132 hot spots for action to address point and nonpoint pollutant sources. These hot spots range from a single industrial facility or factory to an entire municipality or agricultural region. The task force identified 98 "key" hot spots with estimated costs of 8.5 billion ECU. The other 34 spots, all in Denmark, Finland, Germany, and Sweden, were selected by the countries themselves and have an estimated cost of 1.5 billion ECU. Of the 98 "key" hot spots, 47 were identified as priorities, with an estimated cost of 6.5 billion ECU for cleanup and redevelopment. Since the JCP's adoption, 17 hotspots have been removed from the list following the implementation of HELCOM requirements, environmental cleanup, and pollution reduction and mitigation efforts.[35] Because of the JCP's long duration, HELCOM participants plan to revise the plan and the list of hot spots.

In some ways, the JCP is explicitly "post-Westphalian," because it rests on a watershed or hydrographic foundation.[36] Yet, state sovereignty and authority remain apparent. State officials identified and argued over particular hot spots, with the World Bank and several of the wealthiest states in the Baltic region playing active roles in selecting spots in the transition countries. Furthermore, many officials from the wealthier states wanted to avoid identifying domestic hot spots because they felt certain the state would be compelled to clean them up. Identifying these spots at the international level made them domestic priorities. As such, state actors manipulated the resulting list of hot spots.

The HELCOM PITF "initiates, co-ordinates and facilitates the implementation of the Programme."[37] The PITF consists of representatives from all of the contracting parties along with representatives of Belarus, the Czech Republic, Norway, Slovakia, Ukraine, the EBRD, the EIB, the Nordic Environmental Finance Corporation (NEFCO), the NIB, and the World Bank. The International Baltic Sea Fisheries Commission is represented, as are a number of observer organizations including the CCB; the Standing Conference of Rectors, Presidents, and Vice-Chancellors of the European Universities, the European Union for Coastal Conservation, the International Council for Local Environmental Initiatives, the Union of Baltic Cities, and the WWF. The PITF takes an assertive role in formulating, revising, and implementing the JCP. It coordinates international, transnational, and interorganizational cooperation in the development and execution of cleanup and redevelopment programs for hot spots. The PITF appoints lead parties to coordinate specific JCP elements. These parties include states, observer organizations, and other HELCOM standing committees.

JCP implementation requires substantial financial and human resources by all participant states. HELCOM has endeavored to secure support from its wealthier parties and from the European Union (EU), the United States, the four participant multilateral development banks, and other actors in the international development aid community. Nevertheless, the postcommunist states and societies incur substantial costs. The JCP calls for the use of numerous domestic, bilateral, and multilateral funding schemes and identifies a significant private-sector role in the context of privatization, restructuring, and modernization processes underway in the former communist countries. HELCOM and all regime participants have been clear that "co-financing" for environmental projects would be the rule rather than the exception. In other words, local and national beneficiaries of these projects are expected to share costs.

The JCP applies HELCOM's principles and policy norms to international development and restructuring programs within the postcommunist countries of the region, thereby adding economic incentives (or reducing disincentives) for national and local compliance with international standards. In short, the program serves as HELCOM's attempt to organize international environmental and development assistance, as well as domestic environmental and investment programs, within certain ecological and political parameters, thereby combining environmental cleanup with economic development and "transition" assistance. HELCOM "anticipates that national governments, multilateral financial institutions, bilateral organizations and private sector interests will develop projects for implementation during Phase I based on Programme 'priority hot spots.'"[38] The JCP was explicitly formulated to serve as a basis for consideration by the multilateral development banks participating in the task force.[39]

The JCP effectively legitimizes, through the stamp of HELCOM's scientific and technical authority, certain development projects and aid proposals over others and reduces transaction costs for such aid. It constitutes a significant international attempt to integrate environmental considerations into general economic investment planning. JCP participants and documents consistently invoke ecological restoration and pollution prevention, the precautionary principle, the polluter pays principle, best available technology, best environmental practice, information sharing and public release, and the building of national regulatory capacities (such as permitting schemes or emissions control).[40] The program offers a comprehensive outline of steps to improve regional, national, and local ecological quality and state capacity, often addressing specific regions, states, and industrial sectors. In short, the JCP operationalizes the regime's central principles and norms.

By holding out prospects for sharing costs, information, and technology, HELCOM has brought nonlittoral states within the Baltic Sea catchment area into the regime and deepened the commitment of postcommunist states and NGOs to regional cooperation. This change in the incentives for cooperation, in conjunction with the dramatic political changes that took place in the region in the early 1990s, encouraged the postcommunist countries to change positions regarding the detailed pollution information required for the JCP's formulation.[41] The JCP offers wealthier member states opportunities for additional environmental-quality gains in response to domestic political demands, while ensuring that competitive disadvantages are minimized because all other developed littoral states must meet the same standards. Although the ecological benefits of the JCP will be slow to materialize for the Baltic Sea as a whole (because of the slow water-replacement time and the gradual nature of the total pollutant reductions), they could be quite large and occur rapidly in many local areas such as coastal waters, beaches, and rivers. Such improvements may prove economically beneficial and are likely to be politically popular as well.

The implementation of the JCP takes HELCOM deeper into domestic political arenas. It makes HELCOM a player in the reconstruction of postcommunist Estonia, Latvia, Lithuania, and Poland. The list of hot spots, although based largely on the results of scientific research and monitoring, also constitutes an economic development "wish list" for these states. Environmental cleanup and redevelopment programs for each hot spot are assessed by governmental bodies, NGOs, and multilateral banks not only environmentally but also as investments that may or may not yield tangible economic, social, and political benefits.

Co-financing—combining resources from numerous international and domestic sources—has been the primary means of funding most hot-spot mitigation and development programs.[42] International funding sources include the World Bank, the EBRD, the EIB, NEFCO, the NIB, the EU (through its Phare and Life programs),[43] and the WWF, as well as the governments of Finland, Sweden, Denmark, Germany, Switzerland, Canada, the Netherlands, Norway, France, the United Kingdom, and the United States. Of these, bilateral assistance programs on the part of the Western countries of the Baltic region (Denmark, Finland, and Sweden) are by far the largest sources of international funding.[44] Private actors have also invested in the redevelopment of a number of hot spots. Funding programs created and used by donors apply many more criteria in their assessments of various aid proposals than merely determining if a hot spot will be "cleaner" after a specific project than it was before. Social and economic factors associated with economic and political development are considered, as well.

The resulting regional environmental-protection regime has expanded well beyond its original scientific and technological focus. New kinds of expertise and political and ideological discourses associated with post–Cold War Europe, economics, finance, and development now play prominent roles within the regime. Most importantly, the international Baltic environmental-protection regime has become transnational, formulating and implementing a regional set of principles and policy norms for environmental protection within states around the sea. HELCOM moved toward an emphasis on restructuring and reformulating environmental law, policy, and bureaucracy in the postcommunist countries.

Fundamental principles and policy norms institutionalized within international regimes can influence law and regulation in states party to the regime. HELCOM's principles and policy norms are sufficiently flexible, and in some cases vague, to allow for their incorporation into different foreign policies and domestic legal and administrative systems. However, they are not so flexible and vague as to allow incorporation into all such systems. The principles and policy norms considered here, for example, require minimum levels of organizational and administrative capacity, rule of law, and a relative absence of ideological orientations opposed to transnational cooperation and institutional transfer. The "regionalization" of these environmental policy institutions results in greater similarity in the normative and structural foundations of domestic environmental policy in the Baltic region. In this sense, the regionalization of environmental policy is similar (and related) to the regionalization of scientific organizations, research, monitoring, and data gathering witnessed in the region over the last twenty years.[45] The regionalization of environmental policy is part of the larger process of regionalization of shared understandings about relationships between the state, law, and market economies (i.e., the transitions and the large-scale institutional reform and transfer accompanying them).

The transnationalization of principles and norms for environmental policy suggests at least one ecological concern. At the international level, there is scant questioning or analysis regarding the applicability of the transnational principles and norms deemed appropriate at the international level. It is by no means clear that these principles and norms are equally useful and effective, from an environmental standpoint, across countries, environmental problems, and ecosystem types. For example, the regulatory standard of "best available technology" remains problematic in postcommunist countries. Similarly, the "polluter pays" principle relates to different societal conceptions of the role of the state, different ideas about who and what polluters are, and so on. However, the conceptual flexibility of the "polluter pays" principle allows for its incorporation, in very different ways, into domestic environmental policies. For example, some states and societies might choose to implement the principle with sales taxes or another type of value-added tax, whereas others might choose to incorporate it into permitting and licensure systems or a system of tradable permits. Prior research demonstrates HELCOM's role in transferring the "polluter pays" principle and emissions standards as a basis for pollution regulation from western European countries to the eastern and southern Baltic.[46] It is no surprise that transnational principles and norms are more likely to be adopted into domestic law when they are connected to positive incentives such as increased financial resources and improved international reputations.

Environmental law and regulatory policy in the post-Soviet Baltic states are influenced not just by "top-down" pressures from HELCOM, but from "all around." International and domestic organizations, interest groups, and the public may adopt many transnational principles and norms and attempt to hold states accountable to them. For example, bilateral and multilateral criteria for loan and grant programs have been shaped by HELCOM. Increasingly, such programs have participated in the regime's plans and activities, becoming influenced by HELCOM principles, policy norms, and programmatic goals. Thus, HELCOM affects implementation indirectly, through secondary institutions. The reverse is also true: secondary institutions influence these three states through HELCOM. The best example of this is the ongoing HELCOM effort to encourage and facilitate ratification of MARPOL (a collection of international agreements regulating many kinds of marine pollution from shipping) and the implementation of many of the agreements' standards in the Baltic region.[47] In fact, even though the three Baltic states are not all parties to MARPOL, they are participating in HELCOM efforts to implement many of its standards.[48] HELCOM officials recently embarked on a similar mission with regard to hazardous-waste trade regulation, adopting many of the requirements and standards of the Basel Convention on the International Trade in Hazardous Wastes into a HELCOM recommendation in 1998. HELCOM's influence on bilateral and EU assistance programs in the region has also been significant. These programs include implementation of the Helsinki Conventions, HELCOM recommendations, and the JCP in their criteria for funding environmental projects. Special priority is generally given to projects in JCP hot spots and to MARPOL standards. Thus, these assistance programs support the HELCOM environmental policy discourse with resources, providing incentives for states to implement the discourses into policy.

Such overlapping and embedded (or "nested") transnational institutions make the causal pathways between various institutions and organizations extremely difficult to pin down. Challenging methodological questions emerge from such research. The internationalization (or Europeanization) of environmental policy remains difficult to isolate into even a small set of causal relationships. Transnational principles and environmental policy norms, such as those discussed here, are becoming standard in a whole host of international environmental agreements and regimes. As such, determining which one of a number of international influences is primary remains problematic. Furthermore, discursive and other institutional influences are quite diffuse. They affect policy cumulatively, by framing issues and subtly shaping individual and organizational preferences.

The experience of HELCOM demonstrates that international assistance aimed at institutional and organizational capacity building can work.[49] National environmental policymakers' and managers' knowledge and expertise have grown rapidly since the late 1980s. Much of this growth can be attributed to bilateral and international sponsorship of training and assistance programs. Because of lackluster Soviet efforts to implement or publicize HELCOM recommendations, detailed knowledge of them was lacking in the Baltic states.[50] International education and training efforts have helped to rectify this situation. Bilateral and EU assistance programs have also raised the level of knowledge and technical expertise concerning EU environmental policies. HELCOM and EU principles, policy norms, and standards are now making their way into domestic environmental policy in the Baltic states. Furthermore, higher levels of Swedish and Finnish political, economic, and historical involvement with Estonia and Latvia—relative to lesser Danish-Lithuanian cooperation—may partially explain why Lithuania lags behind its Baltic neighbors in environmental policy reform. International support of all three Baltic states remains far below the levels required to fully implement HELCOM's JCP, and the plan remains unimplemented at the majority of hot spots—especially industrial and agricultural ones. However, feasibility studies and mitigation plans for these hot spots are generally complete or underway. Such reports also serve as mechanisms for transnational institutional transfer because they are conducted in accordance with JCP and HELCOM principles and norms.

HELCOM's twenty-five-plus years have engendered trust and familiarity among its many participants. The JCP's regional implementation efforts, projected to last at least twenty years, help to lengthen the time horizons of state, substate, and NGO actors in regional cooperation and increase the appreciation of diffuse reciprocity. This extensive environmental cooperation, coupled with increased regional economic and political openness and integration over time, contributes to improvements in the contractual environment among regional actors. As discussed in the next section, however, much of this environmental cooperation and increased openness remains asymmetrical, as Russia and Belarus lag behind.

Barriers to Cooperation

The principal barriers to environmental cooperation in the Baltic region (and in the Barents region) include (1) historical memory and an associated lack of trust among some parties; (2) differing views regarding nuclear power and weapons; and (3) public-sector capacity in the transition states. That the "Soviet legacy" includes mistrust among members of some ethnic groups and among Russian political elites and those of other postcommunist (and some post-Soviet) elites is not surprising. Many, perhaps most, eastern and central European elites remain deeply skeptical about cooperation of all kinds with Russia. This skepticism is reflected in private conversations, public pronouncements, and media debates. It is identifiable in debate over NATO expansion and citizenship issues vis-à-vis the Russian minorities in the Baltic states. Likewise, Russian officials and political commentators regard the West with clear and perhaps increasing suspicion, fearing that a new line is being drawn across the continent that excludes Russia from Europe.

Certainly, however, historically contingent ideas and biases are not confined to the transition countries. Western officials often remark on the "less European" character of Russian society and politics and the "difficulties" of cooperating with Russians and the Russian state. For Western officials and private-sector actors, operating in the institutional turbulence of the transition countries can be trying. Uncertainty reigns as rules and actors change with remarkable speed, even as much of the inefficiency of the old Soviet-style systems lingers. Westerners regularly encounter forms of "corruption" that are not common in the Western countries of the Baltic region. Russia is often perceived by Westerners as more corrupt than others, though the accuracy of this claim remains uncertain. Lastly, many participants in regional cooperation around the Baltic Sea are quite cynical about the motives of others in the region: Western participants often regard Easterners as interested only in foreign aid and a free ride; Easterners often regard Westerners as arrogant and moralizing in their ultimate pursuit of self-serving goals and interests.

Differing views and domestic policies concerning nuclear power, energy policy, foreign energy dependence, and the utility and morality of nuclear weapons also present challenges to cooperation in the region. As the recent debates about the future of nuclear power in Germany illustrate, these issues remain in flux. The skepticism about nuclear power in Nordic—and even German—popular opinion is not shared among all states and societies in the region. To reduce accident risks, Western states face the need to fund improvements in nuclear safety and disposal in transition countries, when their own domestic policy and publics generally oppose nuclear power and nuclear weaponry. Many Swedish officials, for example, would like Lithuania to commit to closing the Ignalina power plant as a condition for increased international assistance. Norwegian officials and NGOs such as the Bellona Foundation want Russia to agree with them about how and where radioactive materials will be disposed of in Russia. There is also no consensus among state officials in the region over access to information regarding radioactive materials or over the appropriate forms of public participation in these debates. These differences inhibit multilateral cooperation on issues related to radioactive materials.

State skepticism about NGOs remains common among officials all around the Baltic Sea. Russia provides the most extreme case in its prosecution of individuals for cooperating with international environmental NGOs. Whereas international organizations often seek to augment the roles and influence of various NGOs—a post-Westphalian phenomenon—state officials often seek to limit NGO involvement.

Finally, low levels of public-sector capacity in the transition states also impede regional environmental and security cooperation. In Poland and the Baltic states this problem is most apparent at subnational, and particularly local, levels. Although this does not necessarily inhibit interstate cooperation, it certainly inhibits implementation of regional agreements. In Russia and Belarus, the problem is compounded by low and probably declining environmental-policy capacity at the national level. In fact, in May 2000, Putin abolished Russia's agency in charge of environmental protection. Even identifying the appropriate Russian national authorities for some policy areas can be difficult. Russian regulatory authority has been decentralized, and the relationships between national, regional, and local environmental authorities remain unsettled. HELCOM and EU assistance programs attempt to address this problem with some international capacity-building programs, but with little identifiable progress. Russia is by far the largest transition state in the region; as a result, the ability of small international assistance programs and the activities of an environmental organization such as HELCOM to assist or influence Russian policymaking and implementation may be greatly constrained.

Baltic Governance and Regional Peace-Building?

Akin to Conca's notion of "changing the strategic climate," a number of effects of environmental cooperation in the region can be identified. These are most noticeable in the Baltic states, but they also exist in Poland and Russia. Environmental policy elites, tourism officials, port authorities, and urban wastewater-treatment managers know firsthand the benefits of regional cooperation. They express some optimism about the medium- and long-term benefits of environmental cleanup and prevention efforts. Polish and Baltic security and coastal officials benefit from Nordic assistance for marine and aerial surveillance of the coasts. These countries' coast guards have received equipment and training to improve their monitoring of pollution and other activities within their national waters.

Regional environmental cooperation has also helped to change transition-state officials' incentives to cooperate with one another. HELCOM, EU, and NGO programs push for transition-state officials to cooperate. For example, they encourage officials to share experience and expertise regarding the reform and "harmonization" of domestic environmental standards with Western standards. The same is true of internationally sponsored training programs for border and coastal monitoring and control. However, Western officials and organizations have not been consistent on this issue of improved transition-state relations. For example, the EU officially encourages such cooperation, yet it also encourages transition states to compete against one another in their drives for regulatory and market reforms aimed at achieving EU membership.

Regional normative transformation connected with regional environmental cooperation is also evident. Regional ministerial conferences repeatedly promote the notion that the peoples of the Baltic region live in ecological interdependence (as well as in political, economic, and cultural interdependence). Programs to fund curricular change along these lines at primary, secondary, and university education levels aim to put "the Baltic region" at the center of ecological, historical, and cultural education around the region. These programs work explicitly to build or enhance "Baltic identity" and the idea of "Baltic Europe." This conception places geographical, ecological, cultural, and economic factors together under an overarching idea of "Europe." Yet tensions exist within this conception around issues such as the "fit" of Russia and the relative importance of national, regional, and pan-European aspects of collective identity. For example, considerable debate exists in the Baltic states and the Nordic countries over the relative importance of Nordic versus continental European (read: EU) values and institutions. Polish and Baltic-state officials frequently assert the greater "European" character of their states and societies relative to Russia.

Environmental cooperation also contributes to normative changes outside of the environmental arena. For example, interstate and NGO environmental cooperation programs encourage Baltic-state officials and NGOs to attempt to work across domestic ethnic divisions in the Baltic states. International environmental assistance programs encourage multilingual environmental education programs and public-opinion polling. These are difficult issues in the Baltic states, where language requirements for citizenship are common and enforced, and where politicians have few domestic incentives to include Russian speakers—most of whom cannot vote.

The growing number of forums for NGO and subnational regional environmental cooperation suggests that "post-Westphalian" governance is being enhanced on all sides of the Baltic Sea. Such trans-societal relationships became common among Nordic groups and organizations in the period after World War II, though even these levels are being surpassed by the plethora of Baltic regional organizations springing up—most of them funded by Nordic sources. Of particular importance for the construction and spread of common norms are the many NGO-to-NGO links involving environmental advocates, and the many regional professional organizations and associations.

However, a cautionary note about the inclusiveness of emerging ideas concerning who and what constitutes the Baltic region remains important. Russian officials and Russian society are underrepresented in most regional environmental programs. Furthermore, western European (and U.S.) international assistance programs aimed at only those environmental problems in Russia that are perceived as threats to Westerners risk increasing Russian alienation. Norwegian officials and NGOs have made it clear that they are interested in the problems associated with radioactive materials in the Barents because of the threats posed to Norwegians. Many Nordic and HELCOM officials speak of their concern for the pollution emanating from the Russian portion of the Baltic Sea watershed. Similarly, U.S. officials fund programs aimed at dismantling Russian weapons and securing "loose nukes" in Russia because of the threats these pose to Americans. As an aspect of the strategic climate, if Western programs are targeted almost exclusively at things perceived as threats to Westerners, Russian officials will have increased incentives to raise perceptions of threat to Westerners (this dynamic parallels the politics surrounding U.S.-Russian relations and the issue of NATO expansion). Such a dynamic is unlikely to induce regional peace-building or the construction of shared regional identities. Furthermore, it is unlikely to improve the contractual environment between Russian authorities and other participants.

Regional Peace-Building: What Is to Be Done?

If a lasting peace is to be constructed in the Baltic region, three interrelated factors must be addressed: (1) Russian (and Belarusian) inclusion in Baltic and European institutions and transnational communities; (2) the protection of minority rights across the region; and (3) economic stability and growth in Russia. This chapter is concerned largely with the first of these three needs. The second problem, protection of minority rights, remains the subject of many multilateral and domestic efforts, including those of national and international NGOs and intergovernmental organizations.[51] As noted earlier, in the Baltic region a gradual (and still fragile) trend toward greater protection of minority rights appears to be emerging.

Russia and Belarus must be more engaged at all levels and within all kinds of groups if regional peace and greater regional identity and commonality are to be realized. Such engagement costs money. Without more assistance to help build democratic, public-sector capacity in Russia and demonstrate that Western states and peoples care not only about their own interests but also about Russians', multilateral cooperation in the Baltic region is likely to continue to marginalize Russia. The temptation among Westerners to give up on aiding Russia and many other former Soviet republics is high because of high levels of corruption and the repeated failure of many reform efforts. This is particularly true in donor countries such as the United States and the Nordic states, where strong anticorruption norms (as corruption is domestically understood) govern state-society relations and officials' conduct. As noted above, most large international environmental assistance programs are designed to mitigate risks posed to Westerners by problems in the former Soviet Union. This attitude does little to build trust within Russia.

All assistance to Russia and the former Soviet Union should be assessed against its potential to further sustainable democratic development, rather than allowing various international programs and foreign policies to work at cross-purposes. For example, separate U.S.- and NATO-sponsored assistance programs designed to dismantle Russian weapons, enhance civilian control of the military in Russia, and "green" Russian military activities are entirely uncoordinated, giving inconsistent advice and sending mixed messages to the Russian side.[52] Instead of focusing almost exclusively on particular issues seen as problems or threats in the West, greater emphasis must be given to solving Russian problems—that is, those issues of greatest concern to Russians. If Russian elites and publics do not see their concerns and interests reflected in international assistance programs, then regional cooperation is unlikely to yield increases in shared understanding and shared interests vis-à-vis Russia. In other words, without greater efforts to incorporate Russian interests, regional cooperation will have little effect on the strategic climate of Russian cooperation and policymaking or the enhancement of post-Westphalian governance.

More concretely, the creation and maintenance of peaceful regional relations in part requires continued international assistance for democratization and public-sector capacity-building in the Baltic states and central Europe. It requires increases in such assistance to Russia and, if possible, to Belarus. International assistance is not a panacea, of course, and public-sector capacity is not built by merely running a few training workshops and supplying a bit of physical equipment.[53] Public-sector capacity must exist at multiple levels of governance, from the international to the local. Multilateral assistance efforts, environmental and otherwise, must encourage cooperation across and among various levels and groups—East and West, state and nonstate. Such increased contact has the potential to further augment formal and informal regional cooperation, and growing regional identity and awareness of commonality (both interest- and identity-based).

Particular focus is needed in the Baltic region to produce greater equity in transnational relations around the region, so that Russians are not left behind. Clearly, neither national elites nor national polities need to be identified on a par with "the Baltic region." As national collectives, neither Germans nor Russian are likely to see themselves as "Baltic" in the same way as Estonians and Latvians do, for example. Finally, the temptation to single out Russians or transition states as the "patients" in the region should be avoided. Nordic and other Western elites and publics must choose to welcome the peoples and states of the postcommunist countries into the

Baltic region and into Europe. If they do not resist the inclination to leave Russia out of these regional identities and the multilateral institutions associated with them, they risk creating a self-fulfilling prophesy of Russian "otherness"—and regional environmental quality and security are likely to suffer.

Conclusion: Environmental Politics Are High Politics

Environmental issues in the Baltic are no longer "low politics." These issues are not simple or narrowly functionalist areas of cooperation. They command the attention of the region's highest political leaders and they are the targets of significant resource expenditures. At a minimum, it is clear that environmental cooperation around the Baltic Sea has not harmed prospects for regional peace in the region. At present, there is little organized or collective violence in the Baltic and Barents regions, and environmental cooperation is not disturbing this situation. In fact, environmental cooperation remains the main exemplar of regional interstate and trans-societal cooperation. Interstate environmental cooperation, under the auspices of HELCOM, has permeated other international organizations (such as the World Bank), and its content and norms have diffused to the subnational level among governmental and nongovernmental organizations, professional communities, and publics.

Moreover, in all Baltic littoral states—except perhaps Russia—evidence suggests that efforts at environmental cooperation in the region encourage at least small changes in strategic incentives for state officials and publics to engage in peaceful cooperation over the medium and long term. Although such incentive changes are hardly seen at the national level in Russia, they are apparent at the regional level, where officials are often anxious to engage in multilateral cooperation. Changing incentives for interstate cooperation occur in the context of rapidly expanding political, economic, and security-related interdependence of all kinds. Thus, it is difficult to isolate environmentally induced changes in the strategic climate. The same can be said of trans-societal changes. Growing transborder connections among public-sector organizations and individuals and increasing international NGO and private-sector cooperation have all resulted from expanding regional environmental cooperation. Although Russia is less engaged in these changes than are other states in the region, ongoing environmental cooperation between local and regional Russian authorities from northwestern Russia and transboundary NGO linkages vastly exceed those that existed only a few years ago.

The two models discussed in Chapter 1—changing the strategic climate and promoting post-Westphalian governance—appear to work in tandem in the Baltic region. Yet they operate asymmetrically across various Baltic littoral states and societies. Within the empirical world of Baltic regional cooperation, however, the differences between the two pathways can be difficult to sort out. Can strategic incentives be separated from post-Westphalian factors over the medium and long terms? State elites respond to changes in incentives, yet they also play a role in shaping societal transformation. If the character of regional relations changes, so do incentives. If international cooperation influences both drivers of change, it is extremely difficult to separate them in the empirical world. Comparison to other regions, however, suggests that the relatively dense institutionalization of international environmental cooperation in the Baltic region may highlight these difficulties more than in other regions. In other words, in a region with so many avenues for environmental cooperation at multiple levels of governance and among many state and nonstate actors, participants may be responding to changes in both the strategic climate and societal transformation simultaneously.

Implementation of international environmental agreements requires a minimum level of public-sector capacity. International assistance efforts designed to enhance state capacity have borne fruit in the Baltic states and in Poland.[54] These programs provide much-needed additional resources, equipment, skills, expertise, and transnational connection between individuals and groups. They remain centrally important in efforts to enhance environmental protection in the region's transition states. They have the added advantage of working to avoid state failures and, if designed to encourage public participation and social learning, they can incorporate principles and norms of democratic governance in general. Multilateral institutions can contribute to the socialization of newly democratic states and political actors toward more democratic, humanitarian, and environmentally friendly practices.[55] For example, multilateral environmental organizations (both intergovernmental and nongovernmental) have encouraged public participation in environmental policy and have promoted multilingual environmental-education programs, polling, and outreach among various ethnic groups. They attempt to institutionalize public access and participation in domestic and international policymaking in pursuit of more peaceful and democratic resolution of political differences.

Notes

1. See Stacy D. VanDeveer and Geoffrey D. Dabelko, "Redefining Security around the Baltic: Environmental Issues in Regional Context," *Global Governance* 5, no. 2 (April–June 1999): 221–49, from which this introductory section draws. For support of this paper, I would like to thank Ken Conca and Geoffrey Dabelko and the University of Maryland's Harrison Program on the Future Global Agenda and the Woodrow Wilson International Center for Scholars.

2. See Arthur H. Westing, ed., *Comprehensive Security for the Baltic: An Environmental Approach* (London: Sage, 1989); Christian Wellmann, ed., *The Baltic Sea Region: Conflict or Cooperation? Region-Making, Security, Disarmament, and Conversion,* Kiel Peace Research Series no. 1 (Hamburg: Lit, 1992); Pertti Joenniemi, ed., *Cooperation in the Baltic Sea Region* (New York: Taylor & Francis, 1993); Pertti Joenniemi and Peeter Vares, eds., *New Actors on the International Arena: Foreign Policies of the Baltic Countries* (Tampere, Finland: Tampere Peace Research Institute, 1993); Ole Wæver et al., "Societal Security and European Security," in Wæver et al., *Identity, Migration, and the New Security Agenda in Europe* (Boulder: Westview, 1993), 185–99; Olav F. Knudsen and Iver B. Neumann, "Subregional Security Cooperation in the Baltic Sea Area: An Exploratory Study," *NUPI Report* (Norwegian Institute of International Affairs), no. 189 (1995); and Ronald H. Linden, "The Age of Uncertainty: The New Security Environment in Eastern Europe," *Problems of Post-Communism* 43, no. 5 (September/October 1996): 3–14.

3. Unto Vesa, "Political Security in the Baltic Region," in Westing, ed., *Comprehensive Security for the Baltic,* 35–45.

4. Sverre Lodgaard, "Confidence Building in the Baltic Region," in Westing, ed., *Comprehensive Security for the Baltic,* 99–112.

5. Martin O. Heisler, "Migration, International Relations, and the New Europe: Towards Theoretical Perspectives from Institutional Political Sociology," *International Migration Review* 16 (1992): 596–622; and Martin O. Heisler and George Quester, "International Security Structures and the Baltic Region: The Implications of Alternative World-Views," in Olav F. Knudsen, ed., *Stability and Security in the Baltic Sea Region: Russian, Nordic, and European Aspects* (Ilford, U.K.: Frank Cass, 1999).

6. Oran R. Young, "Institutional Linkages in International Society: Polar Perspectives," *Global Governance* 2, no. 1 (1996): 1–24.

7. Ken Conca, "Environmental Cooperation and International Peace," in Paul F. Diehl and Nils Petter Gleditsch, eds., *Environmental Conflict* (Boulder: Westview, 2000).

8. See, for example, Carl Bildt, "The Baltic Litmus Test," *Foreign Affairs* 73, no. 5 (September/October 1994): 72–85; Olli-Pekka Jalonen, ed., *Approaches to European Security in the 1990s* (Tampere, Finland: Tampere Peace Research Institute, 1993); Joenniemi and Vares, eds., *New Actors on the International Arena;* Anatol Lieven, *The Baltic Revolution* (New Haven: Yale University Press, 1993); "Illuminating the 'Gray Zone': A Conference on NATO Enlargement," proceedings of a conference held at the Woodrow Wilson International Center for Scholars, Washington, D.C., December 11–12, 1997.

9. Brian Whitmore, "The Reluctant Dissident," *Transitions* (Prague), May 1998: 68–73; and the Bellona Foundation website: ⟨www.bellona.no/e/rissia/nikitin⟩.

10. "Russia: No Place to Be an Ecologist," *Economist,* June 3, 2000: 52.

11. See, for example, Matthew Bunn et al., "Retooling Russia's Nuclear Cities," *Bulletin of the Atomic Scientists* 54, no. 5 (September/October 1998): 44–50; Matthew Bunn and John Holdren, "Managing Military Uranium and Plutonium in the United States and the Former Soviet Union," *Annual Review of Energy and the Environment* 22 (1997): 403–86; and George Quester, ed., *The Nuclear Challenge in Russia and the New States of Eurasia* (Armonk: M. E. Sharpe, 1995).

12. VanDeveer and Dabelko, "Redefining Security."

13. Interviews conducted in Riga, October 1996.

14. Westing, ed., *Comprehensive Security for the Baltic,* 9. The total number of military vessels in the Baltic Sea has decreased since the late 1980s, given the general regional trend toward declining military spending and the substantial reduction in the naval capacities of the Soviet Union and its successor states.

15. International Council for the Exploration of the Sea (ICES), "Report of the ICES Working Group on Pollution of the Baltic Sea," *ICES Cooperative Research Report,* Series A, no. 15 (February 1970).

16. Ludwik Zmudzinski, "Environmental Quality in the Baltic Region," in Westing, ed., *Comprehensive Security for the Baltic,* 46–51.

17. Martti Koskenniemi, "Environmental Cooperation in the Baltic Region," *Tulane Journal of International and Comparative Law* 59, no. 1 (1993): 85–86.

18. Ibid. This paragraph draws heavily on Zmudzinski, "Environmental Quality," 49–50.

19. See, for example, Medea, *Ocean Dumping of Chemical Munitions: Environmental Effects in Arctic Seas* (McLean, Va.: Medea, 1997); Thomas Nilsen, Igor Kudrik, and Alexandr Nikitin, *The Russian Northern Fleet: Sources of Radioactive Contamination,* Bellona Report no. 2 (Oslo: Bellona Foundation, 1996); Sanoma Lee Kellogg and Elizabeth J. Kirk, eds., *Reducing Wastes from Decommissioned Nuclear Submarines in the Russian Northwest,* proceedings of a NATO Advanced Research Workshop, Kirkenes, Norway, June 24–28, 1996 (Brussels: NATO).

20. This section draws on VanDeveer and Dabelko, "Redefining Security"; and on Stacy D. VanDeveer, "Normative Force: The State, Transnational Norms, and International Environmental Regimes," Ph.D. dissertation, University of Maryland, College Park, 1997.

21. Council of Baltic Sea States, "Presidency Declaration," joint statement from the Baltic Sea States Summit, Visby, Sweden, May 3–4, 1996.

22. Martin O. Heisler and Stacy D. VanDeveer, "The Diffusion of Virtue? International Institutions as Agents of Domestic Regime Change," paper presented at the annual meeting of the Northeast Political Science Association, Philadelphia, November 13–15, 1997.

23. See the website: ⟨www.ee/baltic21/first/htm⟩.

24. Convention on the Protection of the Marine Environment of the Baltic Sea Area, 1974, reprinted in HELCOM, "Intergovernmental Activities in the Framework of the Helsinki Convention, 1974–1994," *Baltic Sea Environmental Proceedings (BSEP),* no. 56 (1994): 107–39 [hereinafter referred to as "1974 Convention"]; and Convention on the Protection of the Marine Environment of the Baltic Sea Area, 1992, reprinted in ibid., 141–85 [hereinafter referred to as "1992 Convention"]. The 1973 Convention on Fishing and Conservation of the Living Resources in the Baltic Sea and Belts (Gdansk Convention) is the other Baltic regional environmental and resource-related agreement.

25. HELCOM's "interim" status ended with the convention's entry into force.

26. See Harold Velner, "Baltic Marine Environment Protection Commission," in Westing, ed., *Comprehensive Security for the Baltic,* 76.

27. Peter M. Haas, "Protecting the Baltic and North Seas," in Peter M. Haas, Robert O. Keohane, and Marc A. Levy, eds., *Institutions for the Earth: Sources of Effective International Environmental Protection* (Cambridge, Mass.: MIT Press, 1993).

28. Velner, "Baltic Marine Environment Protection Commission," 78; Zmudzinski, "Environmental Quality," 49; and HELCOM, "Final Report on the Implementation of the 1988 Ministerial Declaration," *BSEP,* no. 71 (1998).

29. HELCOM, "Final Report."

30. Bengt Broms, "Multilateral Agreements in the Baltic Region," in Westing, ed., *Comprehensive Security for the Baltic,* 64.

31. HELCOM, "Final Report," 1998.

32. HELCOM, "The Baltic Sea Joint Comprehensive Environmental Action Programme," *BSEP,* no. 48 (1993), p. iii.

33. Ibid., p. v.

34. Ibid., pp. vi–vii.

35. Eleven spots (nos. 2, 3, 4, 5, 11, 12, 13, 16, 121, 126, and 131) were pulp and paper mills in Finland and Sweden. Three others are in German cities (nos. 114, 116, and 121), two are in Estonia (nos. 29 and 35), and one is in Russia (no. 68). Therefore, fourteen of the seventeen removed sites are in Western states.

36. I have Matthew Auer to thank for this point.

37. *HELCOM News* 1, no. 1 (March 1995): 2.

38. HELCOM, "Baltic Sea Joint Comprehensive Environmental Action Programme," p. ix.

39. Ibid., ch. 1, p. 2; and HELCOM, "High-Level Conference on Resource Mobilization," proceedings of conference, Gdansk, Poland, March 24–25, 1993, *BSEP,* no. 47 (1993).

40. HELCOM, "Baltic Sea Joint Comprehensive Environmental Action Programme," esp. ch. 5.

41. Ain Laane, "Political Impediments to Implementing International Waterway Agreements: The Example of the Baltic States," paper presented at the annual convention of the International Studies Association, Chicago, February 21–26, 1995, 9–10.

42. Niels-J. Seeberg-Elverfeldt, "'Hot Spotting' in Estonia, Latvia, and Lithuania," *HELCOM News* 1, no. 1 (March 1995).

43. Phare is the main EU program for financial support for restructuring the postcommunist economies of eastern and central Europe. Life is the EU fund for environmental investments within the EU, but some of its money can be spent in states on the EU's periphery.

44. Alexei Roginko, "Domestic Implementation of Baltic Sea Pollution Controls in Russia and the Baltic States," International Institute for Applied Systems Analysis (IIASA) working paper no. 96–91 (Laxenburg, Austria: IIASA, August 1996).

45. Bernt I. Dybern, "The Organizational Pattern of Baltic Marine Science," *Ambio* 9, nos. 3–4 (1980): 187–93.

46. VanDeveer, "Normative Force."

47. Owen Greene, "Implementation Review and the Baltic Sea Region," in David Victor, Kal Raustiala, and Eugene Skolnikoff, eds., *Implementation and Effectiveness of International Environmental Commitments: Theory and Practice* (Cambridge, Mass.: MIT Press, 1998).

48. Owen Greene, oral presentation at IIASA, Laxenburg, Austria, July 1996.

49. Stacy D. VanDeveer. "Changing Course to Protect European Seas: Lessons after 25 Years," *Environment* 42, no. 6 (July/August 2000): 10–26.

50. This was less true in Estonia because of high levels of participation in HELCOM by Estonians during the Soviet period.

51. Examples include Helsinki Watch, Amnesty International, the Organization for Security and Cooperation in Europe, and many bilateral programs. Most of these initiatives support individual human rights advocates and groups on the ground, as well.

52. Stacy D. VanDeveer, "Environment and Security Policy," *Foreign Policy in Focus* 4, no. 2 (1999).

53. On the complexities of capacity building, see Marilee S. Grindle, *Getting Good Government: Capacity Building in the Public Sectors of Developing Countries* (Cambridge, Mass.: Harvard Institute for International Development, 1997).

54. Stacy D. VanDeveer and Geoffrey D. Dabelko, eds., *Protecting Regional Seas: Developing Capacity and Fostering Environmental Cooperation in Europe,* conference proceedings (Washington, D.C.: Woodrow Wilson International Center for Scholars, 2000).

55. Martha Finnemore, *National Interests in International Society* (Ithaca: Cornell University Press, 1996); and Heisler and VanDeveer, "Diffusion of Virtue."

3

Environmental Cooperation in South Asia

Ashok Swain

One-fifth of the world's population and nearly half of the world's poor live in South Asia. The region also shows high rates of population growth. Despite rapid economic growth during the 1990s, the region has among the lowest per capita incomes in the world. Growing subsistence needs impose larger demands on water supplies, arable land, forests, and coastal habitats. Almost all of the seven countries in the region are already suffering from a series of environmental problems in the form of deforestation, soil erosion, and a scarcity of fresh water.

South Asia has long been a region of instability and unrest. It is home to two of the world's most intractable ethnic conflicts: in the Indian state of Kashmir and in Sri Lanka. These two conflicts have generated considerable terrorist violence. The Kashmir conflict has also been the cause of several wars between India and Pakistan—the most recent being the limited war fought in and around Kargil in 1999. The nuclear explosions by India and Pakistan, the two regional giants, in the spring of 1998 have further compli-cated the security environment in the region. A Maoist insurgency in Nepal, a separatist struggle in northeastern India, and a tribal uprising in Bangla-desh are the other serious ongoing conflicts in the region. The failure of the South Asian states to resolve these "traditional" conflicts in the region has prompted many observers to underscore the looming dangers of environ-mental insecurity.

In spite of increasing internal problems and conflicting interests among the countries of the region, however, the prospects for environmental coop-eration still appear to be encouraging. The countries of the region tend to cooperate in combating various environmental problems, particularly water scarcity, through a number of subregional initiatives. India and Pakistan have been sharing the Indus River system for nearly four decades with the help of a bilateral arrangement. On the eastern side, India and Bangladesh have recently signed a long-term treaty to cooperate in using the increasingly scarce waters of the Ganges. India is also increasingly cooperating with Nepal and Bhutan over common water resources. India, Bangladesh, Nepal, and Bhutan formed a subregional group in 1997 to promote specific projects for water and energy resource development.

Though the countries of South Asia have not yet been able to evolve a proper regional machinery to tackle environmental problems, the increasing array of bilateral and multilateral initiatives at the nongovernmental level provides hope for the future. On several occasions, concern for the local environment has also brought people together to challenge the state and its authority. Although these environmental movements have essentially been limited to the local scale, there is a trend to go beyond the local to address issues regionally. Growing interaction among the region's environmental groups has led to several regional environmental initiatives. Given the South Asian response to environmental issues, it can be argued that environmental scarcity not only leads to conflict, but also offers the potential to bring about regional cooperation—even on very inhospitable terrain.

South Asia: A Spawning Ground for Environmental Violence?

Increasingly, a consensus is being reached that there is an environmental dimension to security issues. Though some disagree about its meaning and policy implications, the term "environmental security" is widely used in the academic and policy communities. As Ronnie Lipschutz has written, "from its relatively humble beginning in the early 1980s, the subject of 'environment and security' has become something of a major academic industry."[1] A substantial amount of the research on this topic has been devoted to establishing the environment-conflict linkage—i.e., that environmental destruction may lead to violent conflict.[2] As a result of human-induced environmental destruction, the world is witnessing a sharp reduction in the availability of arable land, forests, fresh water, clean air, and fisheries. The adverse effects of pollution on these scarce resources are worsened by the growing demand for these resources, driven by burgeoning populations, urbanization, conspicuous consumption, and developmental initiatives. The unequal distribution of these resources further complicates the situation. As Thomas Homer-Dixon argues, these sources of scarcity often interact and reinforce one another and lead primarily to two kinds of interaction: *resource capture,* in which pressures of ecological degradation or population growth heighten social conflict over the control of natural resources; and *ecological marginalization,* in which problems of access to natural resources push growing numbers of poor people into practices that worsen ecological stresses.[3] Environmental scarcities, reinforced by resource capture and ecological marginalization, may exacerbate tensions within and between countries. The outcome can be conflict, in the form of civil strife within the nation-state or even "resource war" with other nation-states (although Homer-Dixon found little evidence supporting the latter proposition in the cases he examined).

South Asia is highly vulnerable to this environmental conflict scenario due to its high population density and one of the world's highest levels of poverty. As stated earlier, one-fifth of the world's population and nearly half of the world's poor live in South Asia. In addition, high population growth rates exist throughout the region. A conservative estimate suggests that the South Asian population could grow from today's roughly 1.4 billion to almost 1.8 billion by 2010.[4] India's population officially crossed the 1 billion mark in 2000, and the country adds 17 million people every year. High population growth rates are significantly positively correlated with rates of deforestation and overgrazing and increasing water scarcity in some countries.[5] Researchers refer to the relationship between poverty and environmental issues as the "negative spiral of poverty and degradation," wherein the poor collect as much accessible natural resources as possible to survive, and children are an asset in acquiring these resources, but high fertility contributes to further impoverishment as resources are overexploited.

With massive population growth and rampant poverty in both economic and human terms (see Table 3.1), the countries of South Asia—primarily India, Pakistan, Bangladesh, Nepal, and Bhutan—already suffer from large-scale deforestation, soil erosion, desertification, and depletion of water resources. The scarcity of renewable resources has affected agricultural production, and the natural environment is losing its capacity to support growing human communities. More than ten years ago, Norman Myers made the following prediction about South Asia: "There is a reason to suppose that environmental problems, compounded by population growth, may well undermine the region's economies and disrupt social systems to the extent that they destabilize political regimes and eventually engender

Table 3.1

Human Development Statistics for South Asia

Country	HDI rank[1]	Population, 1998 (millions)	Annual population growth rate, 1998–2015 (estimated, %)	GDP per capita, 1998 (PPP, U.S.$)[1]
Sri Lanka	84	18.5	1.0	2,979
Maldives	89	0.3	2.6	4,083
India	128	982.2	1.2	2,077
Pakistan	135	148.2	2.4	1,715
Bhutan	142	2.0	2.6	1,536
Nepal	144	22.8	2.1	1,157
Bangladesh	146	124.8	1.5	1,361

Source: United Nations Development Programme (UNDP), *Human Development Report 2000* (New York: Oxford University Press, 2000).
[1]For an explanation of the Human Development Index (HDI) and purchasing power parity (PPP), see the notes to Table 2.1 on page 26.

civil disorders culminating in violence—whether violence within individual nations or between two or more nations."[6]

In fact, Myers was quite right when he spoke of environmentally induced violence within nation-states. South Asia has been prone to environmentally induced violence because weak institutions governing the individual states have failed at conflict management. They have witnessed several intrastate violent conflicts over the sharing of natural resources.[7] One clear example is the violence between Tamils and Kannadigas in the southern part of India in 1991 over the water of the Cauvery River. Problems of water sharing have also contributed greatly to violent separatist demands by Sikhs in India and Sindhis in Pakistan. Deforestation and water scarcity have caused local violence in the Himalayan region, as well. Environmentally induced migration has also led to violence in some cases: between Indians and Bangladeshi migrants in India, and between Nepalese migrants and Bhutanese in Bhutan.[8]

The land and topography of the South Asian region are such that it is essentially one geographical unit. The inhabitants of the region also have many civilizational, cultural, and historical commonalities.[9] The region's peoples take pride in their roots in the Indus Valley civilization. Most modern-day South Asian countries were under British colonial rule until the first half of the twentieth century, and nominally independent countries such as Nepal, Bhutan, and Maldives were heavily dependent on the imperial power for their economic survival.[10] Cultural and linguistic linkages have also brought together the peoples of different political units. The reach of Hindi-language movies, Urdu *ghazals* (romantic songs), and Bengali literature is not confined by any particular state boundary within the region.

Complicating matters further, South Asia's national boundaries have no clear lines of demarcation. The postcolonial national territories invariably cut across communities and ethnic groups. People are constantly on the move from one country to another in their quest for survival, adding a new dimension to the already complex political and religious demography of the region. All states in the region, with the exception of India and Sri Lanka, lack a stable political system. Half-hearted efforts to obtain popular legitimacy exacerbate existing tensions in the region.[11] Moreover, the region features two main religious communities, Hindus and Muslims, that have long been antagonistic toward each other. This religion-based confrontation has been a long-standing source of tension and periodic hostilities among India, Pakistan, and Bangladesh, which comprise 94 percent of the region's area and 96 percent of its population.

In short, due to several factors highlighted above, it would seem plausible that the region is highly susceptible to interstate conflict induced by environmental scarcities. However, to date this has not been the case. There are many instances of disputes over the sharing of common natural resources, but the states concerned have not resorted to violent conflict in an effort to resolve them. In many cases, the disputes have ended instead in bilateral agreements. The issue of shared water resources carries the highest potential for violent interstate conflict (with the possible exception of fisheries)—yet shared water resources have actually been the source of extensive bilateral cooperation, even in the face of security crisis. Two major wars, ongoing low-intensity conflicts, and nuclear test explosions have not affected the working of the Indus Waters Treaty. The other major international river system in the region, the Ganges, is also being shared through bilateral arrangements between India and Bangladesh, in spite of the extreme politicization of the issue. Nepal and Bhutan also have several bilateral agreements over the sharing of a common river system with India.

Sharing River Waters Bilaterally

India, Pakistan, and the Indus River

The potential for real military conflict exists between India and Pakistan; the issue of nuclear capabilities has added a further chilling dimension to these countries' tense fifty-year relationship. Surprisingly, however, the Indo-Pakistani dispute over the sharing of the Indus River system has not been as contentious as one might expect.[12] The Indus Waters Treaty of 1960 between India and Pakistan is one of the few examples of successful resolution of a major dispute over an international river basin.

Irrigation along the Indus River is as old as its civilization. A large investment was made by the British colonial administration between 1860 and 1947 to create the Indus basin irrigation system.[13] It is now the largest contiguous irrigation system in the world, with a command area of about 20 million hectares and annual irrigation capacity of more than 12 million hectares. The partition of the Indian subcontinent in 1947 put the headwaters of the basin in India; Pakistan received the lower part of the basin. A serious dispute over these shared water resources occurred in 1948, when India halted water supplies to some Pakistani canals at the start of the summer irrigation season.[14]

The ensuing negotiation between the two countries did not resolve the problem. India's move cut off water to 5.5 percent of Pakistan's irrigated area and put tremendous strain on the new country. The U.S. magazine *Collier's* sent David Lilienthal, one-time chief of the Tennessee Valley Authority, to undertake a fact-finding tour and propose a solution to the dispute. Lilienthal visited the subcontinent and concluded that, even though the two nations quarreled over how much water each received, fully 80 percent of the Indus flowed, unused, into the sea.[15] Lilienthal's article in *Collier's* magazine, "Another 'Korea' in the Making," argued for an early solution to the Indus waters problem and urged that an extended canal system be designed, built, and operated as a unit, jointly financed by India, Pakistan, and the World Bank. Lilienthal's friend and the then-president of the World Bank, Eugene Black, welcomed the idea and offered the bank's support. Although it took nine years for the World Bank to bring the two countries into agreement, the Indus Waters Treaty was finally signed on September 19, 1960, in the Pakistani port city of Karachi.

The approach of the 1960 agreement was to increase the amount of water available to both parties. This future prospect persuaded the two countries to share the quantity of flow and to agree to partition Indus basin waters by allocating the three eastern rivers (the Ravi, the Beas, and the Sutlej) to India, and the three western rivers (the Indus, the Jhelum, and the Chenab) to Pakistan. Partition of the rivers was more acceptable to the countries than joint management, and each country was able to exploit its respective water shares with the help of the Indus Basin Development Fund administered by the World Bank. Detailed provisions were made in the treaty to allow Pakistan to construct a system of irrigation works on the western rivers, to compensate for the loss in irrigation supply from the eastern rivers. The treaty also provided an elaborate system of mutual obligation for the two parties.

The Indus Waters Treaty remains in effect today and has withstood periods of very tense relations between the parties. India and Pakistan have fought a war over an uninhabited area called the Rann of Kutch, and they are still engaged in a low-intensity conflict over control of the Kashmir Valley, but the Indus River arrangement has remained in effect. Perhaps its endurance is due to the fact that the leaders of both countries realize the importance of the Indus waters for their own agricultural production. As Egypt is the gift of the Nile, Punjab is the gift of the Indus. The fertile Punjab region straddles the border between India and Pakistan, giving both countries a strong mutual interest in the proper use of the Indus River.

The geographical setting also facilitated the agreement. It was possible to partition the six rivers in equal numbers between the two countries, and adhering to the principles of partition proved easier than working in a joint management framework. Moreover, the three rivers allotted to Pakistan are relatively unaffected by anthropogenic pollution on the upstream Indian side. The mountainous terrain and ethnic conflicts in Kashmir have restricted population growth and industrial and urban expansion in the upper reaches of the Indus basin. Thus, the water-quality issue has not yet affected the water-sharing arrangement between India and Pakistan, and this condition has helped to limit issues of contention.

The success of the Indus Waters Treaty, then, suggests that if natural resources do not become unmanageably scarce and a strong third-party mediator can promote and facilitate the mutual benefits of cooperation, then there is hope elsewhere for resource-related disputes to be resolved peacefully.

India, Bangladesh, and the Ganges

India and Bangladesh are the two major riparian countries in the eastern Himalayan region. These two neighbors share two large Himalayan river systems, the Ganges and the Brahmaputra. The major focus of dispute between the two countries has been the Ganges, as the Brahmaputra has not yet been substantially tapped.

The dispute over sharing the dry-season flow of the Ganges originated between India and Pakistan in 1951, when India started planning to construct a barrage at Farakka, eighteen kilometers upstream from the border with East Pakistan (later Bangladesh). The plan included a 38-kilometer canal that could divert up to 40,000 cusecs from the dam to supplement the waters of the Bhagirathi-Hooghly. The scheme to divert water from the Ganges to the distributary Bhagirathi-Hooghly was intended to flush out silt and keep the port of Calcutta navigable. The domestic and industrial demands of the city of Calcutta and irrigation schemes in the state of West Bengal also induced the Indian government to develop the project. The Pakistani objection to this plan was very vocal but failed to block India's unilateral decision to begin dam construction in 1962.

The creation of Bangladesh in 1971, although occurring with the active support of Indian forces, did not end this disagreement over the Ganges waters. Various rounds of high-level talks, the formation of the Joint River Commission, and visits by the heads of state of India and Bangladesh could not bring about a permanent solution. The Farakka Barrage became operational for a forty-day trial period in 1975, following a short-term agreement signed by both countries. Bangladeshi president Mujibur Rehman was assassinated in an army coup d'état in August 1975, leading to a deterioration in Bangladesh's relations with its dominant neighbor. Starting in January 1976, India began to divert water unilaterally at Farakka, prompting Bangladesh to raise the issue in various international forums. After a change of regime in India in 1977, the two countries came to a five-year agreement to share the water at Farakka during the dry season. Realizing that the dry-season flow was not sufficient to satisfy their minimum needs, both countries recognized that augmentation of the water supply was the solution to the long-term problem. Both promised to work in this direction.

After the 1977 agreement expired, short-term agreements on dry-season sharing were passed in 1982 and 1985. Since 1988, however, the two countries have been unable to reach agreement due to the decreasing availability of water at Farakka, a result of upstream withdrawals. After years of unsuccessful bilateral negotiation, Bangladesh again raised the issue in several international forums in 1993, which led to a further deterioration in the tenor of bilateral relations.[16]

During the remainder of the year, more than sufficient water flows in the Ganges to meet the demands of both India and Bangladesh. The average minimum runoff at Farakka in the dry season, however, was estimated in 1975 to be only 55,000 cusecs. From this amount, India intended to divert 40,000 cusecs with the help of the diversion canal at Farakka, while Bangladesh demanded all 55,000 cusecs for its own uses. The increasing upstream withdrawal for irrigation purposes in the Indian states of Uttar Pradesh and Bihar further reduced the dry-season flow at Farakka. Beginning in 1994, Bangladesh complained of receiving only 9,000 cusecs in the most acute dry seasons, which indicates that the dry-season water availability at Farakka had dropped to at most 49,000 cusecs.

This new figure created a further hurdle for negotiators.[17] Changes of government in both India and Bangladesh in the summer of 1996 brought new hope of reaching a Ganges water-sharing agreement. The election of Sheikh Hasina, the daughter of Mujibur Rehman, as prime minister of Bangladesh changed the domestic political context. She found that her political interests would be better served by signing a Ganges water-sharing accord with India than by using the dispute as a political weapon. Unlike her predecessors, Prime Minister Hasina was not dependent on India-bashing to acquire legitimacy. Moreover, because of India's unilateral water withdrawal at Farakka in the absence of any agreement since 1988, southwestern Bangladesh was suffering massive losses, mostly in the sectors of agriculture, forestry, and fisheries. For Bangladesh not to find a solution to the water-sharing issue had become economically and environmentally suicidal.

Coinciding with the change of government in Bangladesh, the Indian general elections of May 1996 brought about the defeat of the Congress Party and the introduction of a new coalition government in New Delhi. The United Front government in India and its foreign minister, I. K. Gujral, were interested in living up to their earlier image of friendly relations with their neighbors. The shift marked by the election of Sheikh Hasina in Bangladesh further stimulated the desire in India to improve the bilateral relationship.

In December 1996 the prime ministers of India and Bangladesh signed a new Ganges water-sharing agreement, after a gap of eight years. Instead of their usual short-term approach of sharing the dry-season flow at the Farakka barrage, they agreed to a thirty-year arrangement. The treaty stipulates that when flow rates fall below 70,000 cusecs, India and Bangladesh will each receive half of the water. At flow rates above 75,000 cusecs, India is guaranteed a share of 40,000 cusecs, with the balance of the flow going to Bangladesh.[18]

A striking feature of the agreement is that it overstates the availability of water. The agreement is based on the flow average from 1949 to 1988, but the real flow at Farakka in the 1990s was much lower. Clearly, water experts were aware of the lower flow at Farakka when the agreement was negotiated; thus, politics and diplomacy prevailed over hydrology. In order to make agreement possible, the two countries inflated the figure for available water at Farakka.[19] Although seemingly nonsensical in hydrologic terms, this approach shows the willingness among political elites of both countries to share the Ganges water peacefully.

For Hasina's new government, it would have been suicidal to accept any short-term arrangement in which the Bangladeshi share fell below the amount it was allotted in the 1977 agreement. The position of the Indian prime minister, H. D. Deva Gowda, was also constrained. His United Front government was dependent on the Communist Party of India (Marxist) for its survival, and that party headed the government in the affected state of West Bengal. The Communist leaders would never have accepted an agreement allotting less water to Calcutta. The precarious political situation of the Deva Gowda government would not allow it to reduce India's share of water at Farakka on paper. Moreover, any agreement concerning the sacred Ganges that gave a perceived advantage to predominantly Muslim Bangladesh would have infuriated the opposition party, the Hindu fundamentalist Bharatiya Janata Party. Thus, the manipulation of statistics helped both governments reach an agreement by avoiding immediate internal opposition.

Unfortunately, the very first year of the treaty witnessed a dramatic decrease in the annual rainfall upstream, producing severely low dry-season runoff in the Ganges. With the help of political support, the agreement withstood the decrease, but the rainfall situation has not improved since then. The fluctuation in water flow has now forced both Indian and Bangladeshi authorities to engage in serious negotiation to augment the dry-season runoff. The most important outcome of the 1996 treaty is that it has created a conducive atmosphere for discussing a number of water-related issues between the two countries. The treaty itself also refers to other water-related issues such as flood management, irrigation, river basin development, and hydropower generation for the mutual benefit of the two countries. The signing of the treaty has certainly provided both countries with an opportunity for meaningful cooperation.

India, Nepal, and the Mahakali River

Indo-Nepalese relations have been smooth and friendly in comparison to Indo-Pakistani or Indo-Bangladeshi relations. However, sharing the waters of the major rivers originating in Nepal and flowing into India has strained the bilateral relationship. Negotiations regarding projects on the shared river systems have been dominated by controversies because of a lack of mutual trust.[20] Nepalese rivers have tremendous potential for hydropower generation, but Nepal lacks the capital and technology required for such large projects and also needs a buyer for the surplus hydropower. For various reasons, India is the only country that can provide the needed assistance. Thus, India's direct involvement in the use of the river water has been crucial. In general, however, the Nepalese feel that they have not been treated equitably under the various water-resource development agreements with India, including those governing the waters of the Sarada (1920), the Kosi (1954), and the Gandak (1959) rivers.[21]

In the 1990s, development projects on the Mahakali River raised a serious controversy between India and Nepal. In December of 1991, both countries signed a memorandum of understanding (MOU) to construct a barrage at Tanakpur, for which Nepal agreed to provide 2.9 hectares of land. This issue became extremely controversial in Nepal because of the country's domestic political situation. The Nepalese opposition framed the issue in nationalist terms, raising concerns about Nepal's territorial sovereignty. Opposition leaders argued that while signing the agreement, the government of Prime Minister G. P. Koirala had overlooked Nepal's interests in order to appease India. India, at the same time, resisted any change in the agreement on Tanakpur.

Although the government of Nepal at the time attempted to frame the MOU as an "understanding" and thus a nonconstitutional issue, the opposition demanded parliamentary ratification of what was argued to be a "treaty" involving the sharing of the country's natural resources.[22] The MOU seemed to imply that India had given Nepal electricity and water for irrigation as a friendly gesture, but nationalists in Nepal argued that these benefits were actually the country's rightful payment for having conceded its territory to complete the project.

The validity of the MOU provisions was challenged in Nepal's Supreme Court. The plaintiffs sought a ruling that the MOU had to be submitted for ratification in accordance with the country's constitution. In December 1992 the court decided that the MOU was in fact a treaty, necessitating ratification, but it also allowed the government to decide whether the ratification was to be accomplished by a two-thirds majority in a joint session of parliament or by a simple majority in the lower house. The government subsequently formed a committee to evaluate the impact of the project from social, political, and diplomatic perspectives. It concluded that the impact of the treaty was not of a "pervasive, long-term, and extensive" nature. Following these findings, the government sought to ratify the treaty using a simple majority vote. Both the opposition and elements within the ruling Nepali Congress Party rejected the simple-majority motion, and an all-party parliamentary committee was formed to decide what was to be done.

Subsequently, the Koirala government announced midterm elections, which were held in December of 1994. The Nepali Congress Party lost in the election; the Communist Party of Nepal (UML) emerged as the single largest party and formed a minority government. After nine months in power, the UML itself was voted out and replaced by a coalition of the Nepali Congress Party and the Rastriya Prajatantra Party, with Sher Bahadur Deuba of the Nepali Congress Party as prime minister. In December 1995, the Indian foreign minister visited Kathmandu and negotiated the Treaty on Integrated Development of the Mahakali River with his Nepali counterpart. This treaty stipulated that the Detailed Project Report on the dam be completed six months after an exchange of the instruments of ratification, that financing be organized within one year, and that construction be completed within eight years. On February 12, 1996, the prime ministers of India and Nepal signed the Mahakali treaty. The treaty was subsequently ratified by the Nepali parliament as per the country's constitution, and in June 1997, both countries exchanged the instruments of ratification of the treaty, which will remain valid for seventy-five years.[23]

This treaty emerged as a solution to the legacy of disagreement between Nepal and India over the Tanakpur Barrage project. The treaty brought three separate water-resource projects under its ambit. In addition to validating the Tanakpur MOU, the Mahakali treaty took under its wing a regime established seven decades earlier by the Sarada treaty. The Tanakpur Barrage is linked with the Sarada Barrage, which was built in the 1920s after an agreement between British India and Nepal to exchange 4,000 acres of territory at the eastern flank of western Nepal. As Salman Salman and Kishor Uprety argue, "the Mahakali treaty is a first in many ways. It lays down the principle that as a boundary river on large stretches, the Mahakali will be developed in an integrated way to maximize the total net benefit from development. Both parties will, in theory, be entitled to equal benefits, and will share the costs in proportion to the share of benefits they actually receive."[24]

The Mahakali treaty also paved the way for the construction of the Pancheshwar multipurpose project.[25] The agreement set the basis for developing the 6,480 megawatt Pancheshwar hydropower project on a stretch of the Mahakali that crosses the border between the two countries. When completed, the 315-meter-high, rock-fill structure will be the second highest dam in the world (after the Rogun Dam in Ukraine). The 12-billion-cubic-meter reservoir created will store the flow of the Mahakali River to yield a discharge of 885 cubic meters per second below the powerhouse, which will produce up to 3,240 megawatts of power on each side of the river. The reservoir will inundate 134 square kilometers of land (54 km^2 in Nepal and 80 km^2 in India) and is likely to directly displace 65,000 people (15,000 in Nepal and 50,000 in India).The Mahakali treaty is undoubtedly a very positive step toward further bilateral cooperation on water-resource development between India and Nepal (although controversies around the agreement endure, as discussed below). Both sides have gradually come to realize that their own interests are better served through mutual coopera-tion. Nepal's vast hydropower potential and regulated releases of water therefrom to Indian agricultural fields could immensely benefit both coun-tries.[26] The treaty has also enabled the establishment of the Mahakali River Commission, which has a relatively broad mandate. The comprehensive nature of the treaty shows the determination of India and Nepal to move beyond the unilateral water-development strategy to which they had pre-viously adhered and instead promote and strengthen their bilateral coopera-tion in order to receive the maximum benefits of shared river waters.

Establishing Regional Cooperation for Environmental Management

The challenge that lies before South Asian states is to provide their growing populations with access to clean drinking water, reliable irrigation, cheaper energy resources, and flood protection. As the above-mentioned cases demonstrate, there are signs of growing bilateral cooperation among gov-ernments in South Asia to share and develop water resources. This coopera-tion has been achieved in spite of strongly antagonistic historical, ethnic, and political factors. Increasing water scarcity and its direct, adverse effect on the agricultural sector have forced the authorities to give priority to cooperative water management. Yet some environmental groups criticize these bilateral attempts to harness the river systems because of their em-phasis on large-scale water infrastructure projects. The Mahakali treaty between India and Nepal has been particularly subjected to this criticism.[27]

It fails to satisfy environmentalists, as it has not sincerely addressed the social and environmental effects of the various construction projects it comprises, and it has not involved ordinary people in the management of a shared water resource.[28]

To address these challenges, water agreements must be innovative and look beyond the conventional path of development. These initiatives must strive to build institutional responses to meet water and energy demand with minimum adverse social and environmental costs. Moreover, the water-scarcity issue cannot be addressed adequately by working only on the supply side. For fruitful and long-lasting cooperation on shared water resources, a comprehensive approach to the water scarcity issue is needed. Such an approach requires a series of measures to be taken at the basin level, focusing on watershed management, water quality, and land-water interaction. Management of shared river systems must grow beyond the sphere of national sovereignty and bilateral cooperation; it must be addressed at the regional level to achieve the best possible use of available water.

Moreover, as suggested previously, South Asia suffers heavily from other forms of environmental degradation. As with water, issues such as deforestation in the Himalayas, air pollution from thermal power production, and pollution of the coastal environment also demand regional action; they cannot be addressed effectively by individual states. Thus, the South Asian region has an urgent need to develop a comprehensive regional program of cooperation for the protection and management of the shared environment.

In addition to occasional regional input into the development of institutional, legislative, and technological frameworks at the national level, there have been some recent attempts at strengthening regional cooperation through the formulation of a number of regional environmental programs.[29] On the initiative of the United Nations Development Programme, all of the South Asian countries, as well as Afghanistan and Iran, came together in 1980 to establish the South Asian Cooperative Environment Program (SACEP). At their first meeting in Bangalore, India, the member countries identified the broad areas in which immediate cooperation was required; various countries agreed to be the focal point for each of these areas. The SACEP Secretariat was established in Colombo, Sri Lanka, in 1981 to coordinate program activities. In 1984, SACEP initiated a Regional Seas Program for the South Asian seas and also decided to undertake urgent cooperative action to resist deforestation and promote reforestation schemes in the region. SACEP planned the "Year of Trees for South Asia" for 1988.[30] Unfortunately, due to the escalating ethnic crisis in Sri Lanka, the work of SACEP was brought to a near standstill from 1984 until the early 1990s. SACEP revived itself in the 1990s and approved an action plan called SACEP's Strategy and Program. This program's key activities focused on capacity building, awareness raising, information exchange, technology transfer, environmental-management and institutional-development training, and ecosystem, watershed, coastal resource, and wildlife management. However, SACEP's Strategy and Program failed to provide any significant contribution to regional environmental protection during its tenure (1992–96), quite possibly because it focused on too many issues at once.

In 1985, seven South Asian countries officially formed a regional organization, the South Asian Association for Regional Cooperation (SAARC), to promote cooperation in science, education, technology, cultural exchanges, and other matters of mutual concern and interest. At the third meeting of SAARC heads of state, held in 1987, it was decided to study and then issue recommendations on how regional cooperation could respond to rapid and continuing environmental degradation. A 1988 meeting of SAARC country experts identified common areas of regional concern and measures required to address those concerns. Recommendations were made concerning the exchange of information and regional planning for the appropriate use of land and water resources. Since South Asia houses some of the world's largest river systems, the experts group suggested a program of integrated development of river basins. At its fifth summit in Male, the capital of Maldives, in 1990, SAARC declared 1992 to be the "SAARC Year of the Environment."

Undoubtedly, one of the major developments in the past decade has been the recognition by South Asian countries of the importance of environmental concerns and the need for cooperative efforts. Unfortunately, however, this recognition has not yet resulted in concrete regional action. Several factors have hindered successful and effective regional initiatives for better management of the environment in South Asia.

The first factor inhibiting cooperation is the lack of a strong regional environmental agency. SAARC is not yet as powerful an organization as the Association of Southeast Asian Nations or the Southern African Development Community (SADC). SAARC has a number of inherent contradictions to resolve before it becomes a strong regional mechanism. From its inception, SAARC has been plagued by internal tensions, which can be traced to geopolitical imbalances among the member countries. Perhaps no other geographical region is dominated by a single power to the extent that South Asia is dominated by India. India has a population three times as large as the combined populations of the six other South Asian countries, it occupies 73 percent of the total area of the region, and its gross national product is three-fourths of the region's total.[31] India is not only the region's largest and strongest country, but it also constitutes the core of the region.[32] The fear of India by its smaller neighbors leads them to forge extraregional connections to blunt the edge of India's domination, efforts that have reduced the effectiveness of SAARC. The extremely bitter relationship between the two major member states, India and Pakistan, further reduces SAARC's importance.

A second factor inhibiting cooperation involves the complex political terrain of South Asia. Ruling elites in most of the South Asian countries are not enthusiastic about strengthening regional ties. Although India and Sri Lanka have been able to develop and sustain democratic systems, Bangladesh is working hard to cope with its newly found democratic structure, and Pakistan has returned to military rule. Nepal has recently become democratic, but Bhutan remains under the rule of a traditional monarch and the small island nation of Maldives is governed by a one-party system. So although the ruling elites of South Asia do show concern about regional environmental problems, they lack the necessary political will, strength, and maturity to entrust such problems to a supranational body such as a regional organization.

Primarily because of a lack of unity and commitment to a stronger regional arrangement among its member states, SAARC has remained a mere meeting and discussion club. All of the South Asian nations must realize that their ultimate self-interest is inevitably merged in the inescapable web of interdependence.[33] As this understanding has so far been lacking, SAARC has failed to find meaningful regional cooperation on environmental issues.

The stalemate in SAARC has led to attempts to create a subregional organization to deal with the most important and sensitive environmental resource, shared river systems. Four SAARC countries—India, Nepal, Bhutan, and Bangladesh—decided in 1996 to establish a "growth quadrangle."[34] (These countries are the riparian nations of all the eastern-flowing Himalayan rivers.) The increasing scarcity of water and energy has forced these countries to realize that they need to work on specific cooperative projects in the water and energy sectors. They realize that the appropriate development of shared water resources will necessitate basin-based cooperative efforts. The idea to bring all the riparian nations under a subregional outfit was formally mooted by Bangladesh in December 1996; the proposal was immediately supported by Bhutan and Nepal. India, though initially reluctant, finally decided to back the initiative.[35]

The growth quadrangle is an excellent initiative, but because of objections raised by Pakistan, it has become bogged down in the legalities of the SAARC Charter and procedures.[36] Pakistan perceives the initiative as targeted at isolating Pakistan within SAARC, and Islamabad has therefore reacted unfavorably. Sri Lanka and Maldives were also unhappy at being left out of the initiative. To further isolate Pakistan, India requested in 1997 that Sri Lanka coordinate subregional cooperation among India, Sri Lanka, and Maldives in several sectors, particularly trade, tourism, and fisheries. Success in establishing these subregional groupings will certainly influence cooperation on common environmental issues. However, powerful groups, particularly in Bangladesh, Nepal, and Sri Lanka, are generally opposed to any subregional initiatives due to their fear of complete Indian domination of any grouping that does not include Pakistan.

The continuing political problems among the member states are a major obstacle to the successful operation of SAARC. Most of the countries in the region continue to project their bilateral differences and conflicts to achieve short-term political ends, a pattern that seriously affects the evolution of a fruitful regional cooperation mechanism. Moreover, the SAARC member states have almost completely ignored the involvement of nonstate actors in designing and executing regional policies.[37]

Hope for the Future: Moving beyond Boundaries

The end of the Cold War has seen the emergence of people-to-people dialogue and exchanges in many parts of the world. These unofficial dialogues and related training and exchange programs cover a wide range of issues, including environmental cooperation; they sometimes supplement official interstate relations and sometimes even become an alternative approach to cooperation. Yet in South Asia, these "track two" processes have so far had an insignificant influence on regional politics, for they face a number of serious hurdles at the interstate level. In particular, government impediments create an enormous communication and information gap between the people of India and Pakistan.

Unofficial dialogues and exchanges are gradually being seen as a useful tool to address regional problems, however. The growing presence of democratic space in much of the region encourages wider public involvement in regional issues—in spite of resistance from state officials and, in some cases, from the military-bureaucratic oligarchy. In keeping with the South Asian tradition, debate rages about the usefulness of these unofficial exchanges of ideas. Some argue that this approach is ineffective without the support of governments. Others, however, see these unofficial initiatives as crucial to the creation of a conducive atmosphere for regional cooperation because the political institutions that might lead the way are fragile and weak.[38]

The process is not limited to debates; some concrete efforts have been made in recent years within the unofficial sectors to start a dialogue at both the bilateral and multilateral levels. Most of these nonstate initiatives are trying to address several issues, with the main focus being on security. The Kashmir issue, for example, has been the major focus of unofficial interactions between India and Pakistan.[39] Recently, track-two diplomacy has also been advocating nuclear restraint. Unofficial dialogue in the region also takes place regularly on cross-border terrorism, illegal migration, and trade issues. Nevertheless, environmental problems have gradually begun to gain the attention of these groups. The encouraging news is that a growing number of unofficial meetings between South Asian policymakers and opinion makers are showing interest in discussing the possibility of regional or bilateral cooperation on environmental issues.

Bilaterally, unofficial initiatives related to the environment have concentrated on shared river management. This is not surprising, as it is the most important environmental issue affecting the bilateral relations of most South Asian countries. The Patna Initiative aims at interaction between academics and scientists in the Ganges sub-basin shared by India and Nepal. This initiative was started in 1992 by two research institutes, the Nepal Water Conservation Foundation and the Center for Water Resources, located in Bihar, India. With the help of journalists, this initiative has issued several research articles and recommendation papers on the Koshi project on the Ganges.[40] The Patna Initiative tries to identify and develop subregional water-management strategies, particularly at the local level.

The Rajiv Gandhi Institute for Contemporary Studies, a highly influential New Delhi think tank, has taken the initiative since the mid-1990s in organizing bilateral dialogues between India and Nepal, India and Sri Lanka, India and Pakistan, and India and Bangladesh. Academics, journalists, and retired civil servants and diplomats are encouraged by this organization to come together in order to develop a shared perspective on issues of common benefit. Among other issues, the Indo-Nepal and Indo-Bangladesh dialogues have tried to find long-term solutions to the sharing and management of common water resources.

The Center for Policy Research in New Delhi and the Center for Policy Dialogue in Dhaka have also organized Indo-Bangladeshi dialogues since 1994, with the help of the Ford Foundation. This forum helps politicians, academics, journalists, and retired diplomats from both countries to meet in order to improve bilateral relations. Finding ways to solve disagreements over shared river resources is one of the forum's four identified areas of discussion.

In addition to bilateral initiatives, some unofficial multilateral initiatives also specifically focus on river-water management. The Eastern Himalayan River Study is a forum for policy research centers in India, Nepal, and Bangladesh, supported with both internal and external funding. The Center for Policy Research in Delhi, the Institute for Integrated Development Studies in Kathmandu, and the Bangladesh Unnayan Parishad in Dhaka have taken up the task of finding common approaches to water-resource development projects and water sharing in the Ganges-Brahmaputra basin.

At present only a few unofficial initiatives are being undertaken at the regional level to address other, non-water-related environmental issues. One is the Climate Action Network of South Asia (CANSA). Established by the Bangladesh Center for Advanced Studies, CANSA spreads information about greenhouse-gas problems in the region in collaboration with independent research institutes in other countries of South Asia. The Center for Science and Environment, an Indian environmental research and activist organization, has been involved since 1989 in periodically organizing meetings of South Asian environmentalists to develop a common regional position on environmental issues. Recently, the Colombo-based Regional Center for Strategic Studies and the Sustainable Development Policy Institute in Islamabad have organized workshops and sponsored research on regional environmental issues. A recent study by a group of South Asian scholars on the cost of noncooperation, using different sectoral approaches, has shown that cooperation is not only environmentally sound but also cost-effective.[41] The sharing of ideas by these groups can provide forums for the beginning of regional processes of resource sharing.

For a variety of reasons, nonstate actors in South Asia have had very little influence over the formulation of their countries' environmental policies, as their countries' respective bureaucracies have traditionally monopolized the process. The region has seen considerable growth in the number of environmentally focused nongovernmental organizations (NGOs) over the last twenty-five years. However, the vast majority of them are very small and their influence is limited to the grassroots level. Because of massive domestic environmental problems, most of these groups have concentrated their efforts on the domestic situation and ignored the government's regional or global environmental policy.[42] Very few NGOs in South Asia possess the resources or the interest to influence their country's regional environmental policy.

In recent years, dialogue at the unofficial level has generated interest among elites; academics and the media are increasingly recognizing their own importance. However, these interactions and deliberations have not yet successfully dispelled the distrust that exists among the ruling establishments of the South Asian countries. Thus, they have been unable to produce any breakthroughs on contentious regional issues, including environmental ones. To bring about meaningful regional cooperation on environmental issues, policymakers' perceptions need to be changed. This change can be achieved only with increased popular interest and participation. Unfortunately, until recently, environmental consciousness among the South Asian masses has generally been low.[43] Since the beginning of the 1990s, however, popular awareness and involvement on some environmental issues has been growing. As Richard Matthew has observed, "pushed into higher and higher levels of vulnerability, it may be the people of South Asia who will have the incentive and need to rethink social practices, values, institutions and beliefs in the context of global environmental change and local community, thereby developing new environmentally sustainable approaches to political and economic life."[44]

Indian citizens are increasingly organizing themselves to protest, mobilize public participation, and create associations that can take action to protect their environment. In recent years, forest-based popular protests have successfully asserted people's right to manage forests and brought about a significant change in the existing forest policy in order to facilitate public involvement. After achieving some success regarding their demands in the forest sector, the focus of the environmental movements in India has been redirected against the building of large dams.[45] Bangladesh has also recently witnessed successful environmental protests against proposed massive projects for flood management.[46] In Nepal, the authorities are facing serious public opposition to their attempts to build dams. However, these popular actions and environmental pressure groups have mainly been influenced by the "NIMBY" (not in my back yard) syndrome, in that they resist population-displacing development projects.

In recent years, however, environmental movements starting from locally based opposition to individual development projects are moving toward more encompassing issues and campaigns. The spread of education and information has led to the gradual diffusion of the popular interest in and action to address regional environmental issues such as air and water pollution and protection of biodiversity. The interaction among environmental groups of the region is also increasingly highlighting the fruits of regional initiatives. These developments all have the potential to gradually pressure the governments of South Asia into environmentally sensitive policymaking and cooperation with other countries to effectively address environmental threats.

Conclusion

Most environmental issues are transboundary in nature. Not only will they gradually force the countries of South Asia to adopt a regional cooperative approach, they may also help them to put aside extremely politicized and highly emotional issues of contention. It will be relatively easier to initiate cooperation on environmental matters than on other issues; environmental cooperation may in turn spill over to help the countries in the region build the mutual trust necessary to address other traditional issues of dispute. The contribution of the 1996 Ganges agreement between India and Bangladesh is a good example in this regard. Following the agreement, both countries are trying to sort out their differences over territory, trade, and transit in an atmosphere relatively free of hostility. The 1996 Mahakali treaty has also brought several positive developments in the bilateral relationship between India and Nepal. In December of that same year, the Indo-Nepali Treaty of Trade was renewed for a period of five years, with the provision for automatic renewal every five years thereafter. Under the Treaty of Trade, India provides duty-free access to the Indian market, on a nonreciprocal basis and without quantitative restrictions, for most articles manufactured in Nepal. In spite of several political changes in Nepal and India, recent years have witnessed bilateral contacts at practically all levels.

This spillover effect has not been produced, however, by the 1960 Indus Waters Treaty between India and Pakistan. As with the Inguri River in

Georgia,[47] Indo-Pakistani cooperation over water resources has been maintained while the riparian parties are virtually at war with each other. Cooperation on the Indus River shows that even under the most problematic conditions, water can bring cooperation. However, it is up to the state actors to make choices and nurture that water-based cooperation into peacemaking. Unfortunately, India and Pakistan have not yet made that choice. Increasing water scarcity might force them to come together in the near future to improve their cooperation on the Indus River system and to adopt integrated water development in the basin. The need for water certainly has the potential to push the Kashmir issue off the agenda for some time and force the region's two most powerful countries to cooperate. It would certainly be a great boost for South Asian regional cooperation to emerge from the malignant shadow of the antagonistic India-Pakistan relationship.

Post-Westphalian rationality advocates that the nation-state has lost its historical usefulness, and that, as a result, solutions to the problems of security and welfare need to be located in international or regional structures.[48] The Westphalian order, inaugurated with the Peace of Westphalia in 1648, brought about the state-formation and nation-building process seen in Europe for more than 350 years. The South Asian states, however, subjects of colonialism until the mid-twentieth century, began their nation-state projects only in the last fifty years, and their process is far from complete. Their premature exposure to the crisis of the Westphalian order has brought hesitance and confusion among state elites forced to adapt to the changing international situation. In spite of active opposition from some powerful sectors in these still-strong states, a regional identity is increasingly emerging. Internal awareness and external encouragement are strengthening this evolutionary process.

Transboundary environmental issues are gradually beginning to challenge the sacred boundaries of national sovereignty in South Asia. A nation-state alone is not capable of solving many of the environmental problems that it faces. The sharing of international river water, declining fish catches in the open sea, and increasing air pollution have exposed the hollowness in the authority of an individual state to find solutions. The realization that many environmental issues require genuinely regional action has prompted civil society groups in the region to exert pressure on the reluctant states to come closer and work together. Greater regional environmental cooperation in the future can ensure further economic development, infuse greater political stability, and enhance human security.

Notes

1. Ronnie D. Lipschutz, "Environmental Security and Environmental Determinism: The Relative Importance of Social and Natural Factors," paper presented at the North Atlantic Treaty Organization Advanced Research Workshop titled "Conflict and the Environment," Bolkesjø, Norway, June 12–16, 1996.

2. Some of the research that has theoretically or empirically tried to establish the link between environmentally induced resource scarcity and conflicts in society are *Environment and Conflict,* Earthscan briefing document no. 40 (London: Earthscan, November 1984); Thomas F. Homer-Dixon, *Environment, Scarcity, and Violence* (Princeton: Princeton University Press, 1999); Peter Wallensteen, "Environmental Destruction and Serious Social Conflict: Developing a Research Design," International Peace Research Institute [PRIO] report no. 3 (Oslo: PRIO, May 1992): 47–54; Ashok Swain, *Environment and Conflict: Analysing the Developing World,* report no. 37 (Uppsala: Department of Peace and Conflict Research, Uppsala University, 1993); Ashok Swain, *The Environmental Trap: The Ganges River Diversion, Bangladeshi Migration, and Conflicts in India,* report no. 41 (Uppsala: Department of Peace and Conflict Research, Uppsala University, 1996); Nina Græger and Dan Smith, eds., *Environment, Poverty, Conflict,* PRIO report no. 2 (Oslo: PRIO, 1994); Nils Petter Gleditsch, ed., *Conflict and the Environment* (Dordrecht: Kluwer, 1997); and Günther Bächler, *Violence through Environmental Discrimination* (Dordrecht: Kluwer, 1999).

3. Homer-Dixon, *Environment, Scarcity, and Violence,* 177.

4. Sandy Gordon, "Resources and Instability in South Asia," *Survival* 35, no. 2 (Summer 1993): 66–87.

5. UN Environment Programme (UNEP), *Global Environmental Outlook 1: Global State of the Environment Report* (Nairobi: UNEP, 1997); and UN Economic and Social Commission for Asia and the Pacific (ESCAP), *State of the Environment in Asia-Pacific 1995* (Bangkok: ESCAP, 1995).

6. Norman Myers, "Environmental Security: The Case of the Indian Sub-Continent," paper presented at the Pugwash conference, Dagomys, Soviet Union, August 29–September 3, 1988.

7. Ashok Swain, "Fight for the Last Drop: Inter-state River Disputes in India," *Contemporary South Asia* 7, no. 2 (July 1998):167–80.

8. Ashok Swain, "Environmental Migration and Conflict Dynamics: Focus on Developing Regions," *Third World Quarterly* 17, no. 5 (December 1996): 959–73; and Ashok Swain, "Displacing the Conflict: Environmental Destruction in Bangladesh and Ethnic Conflict in India," *Journal of Peace Research* 33, no. 2 (May 1996): 189–204.

9. Partha S. Ghosh, *Cooperation and Conflict in South Asia* (Dhaka: University Press, 1989).

10. Khem Kumar Aryal, "Finding South Asia," *Journal of Peace Studies* 7, no. 2 (March–April 2000): 46–49.

11. Lok Raj Baral, "SARC, but No 'SHARK': South Asian Regional Cooperation in Perspective," *Pacific Affairs* 58, no. 3 (Fall 1985): 411–26.

12. Robie I. Samanta Roy, "Remote Sensing in South Asia for Water Resource Management and Conflict Resolution," Institute for Defense Analyses, Washington, D.C., 1998.

13. Syed Naseer A. Gillani and Mohammed Azam, "Indus River: Past, Present and Future," in Aly M. Shady et al., eds., *Management and Development of Major Rivers* (Calcutta: Oxford University Press, 1996).

14. G. T. Keith Pitman, "The Role of the World Bank in Enhancing Cooperation and Resolving Conflict on International Watercourses: The Case of the Indus Basin," in Salman M.A. Salman and Laurence Boisson de Chazournes, eds., *International Watercourses: Enhancing Cooperation and Managing Conflict,* World Bank technical paper no. 414 (Washington, D.C.: World Bank, June 1998).

15. Dante A. Caponera, "International Water Resources Law in the Indus Basin," paper presented at the regional symposium titled "Water Resources Policy in Agro-Socio-Economic Development," Dhaka, August 4–8, 1985.

16. For a detailed description of the conflict, see Ashok Swain, "Conflicts over Water: The Ganges River Dispute," *Security Dialogue* 24, no. 4 (December 1993): 429–39.

17. Swain, *The Environmental Trap,* 48.

18. Treaty between the Government of the Republic of India and the Government of the People's Republic of Bangladesh on Sharing of the Ganga/Ganges Waters at Farakka (hereinafter Ganges River Agreement), December 12, 1996.

19. Ashok Swain, "Reconciling Disputes and Treaties: Water Development and Management in Ganga Basin," *Water Nepal* 6, no. 1 (1998): 43–65.

20. Salman M.A. Salman and Kishor Uprety, "Hydro-Politics in South Asia: A Comparative Analysis of the Mahakali and the Ganges Treaties," *Natural Resources Journal* 39, no. 2 (Spring 1999): 295–343.

21. B. C. Upreti, *Politics of Himalayan River Waters: An Analysis of the River Water Issues of Nepal, India, and Bangladesh* (Jaipur, India: Nirala, 1993).

22. The Nepali Constitution, article 126, stipulates that the parliament must ratify any agreement on the sharing of the country's natural resources that has "long term, pervasive and serious" impacts on the country.

23. *Deccan Herald,* June 6, 1997.

24. Salman and Uprety, "Hydro-Politics in South Asia," 312.

25. Swain, "Reconciling Disputes."

26. B. G. Verghese, "Give the Gujral Doctrine a Chance," Rediff on the Net, ⟨www.rediff.com⟩, 1997.

27. The Indus Waters Treaty was concluded at a time when large projects were the panacea of development and growth, and the 1996 Ganges agreement is primarily a water-sharing agreement; it does not have any specific plan for large-scale projects.

28. Environmental groups and activists in South Asia regularly blame their "quasi-states" and interventionist external forces for the degradation of the natural environment, large-scale displacement of the population, and the failure to achieve development. See Nauman Naqvi, ed., *Rethinking Security, Rethinking Development: Anthology of Papers from the Third Annual South Asian NGO Summit* (Islamabad: Sustainable Development Policy Institute, 1996).

29. R. B. Jain, "Conflict and Cooperation on Environmental Issues in South Asia," paper presented at the international seminar titled "South Asia at the Crossroads: Conflict and Cooperation," Bangladesh Institute of International Strategic Studies, Dhaka, February 6–8, 1994.

30. SACEP, *Newsletter* 2 (July 1984).

31. Emajuddin Ahamed, *SAARC: Seeds of Harmony* (Dhaka: University Press, 1985).

32. K. Raman Pillai, "Tensions within Regional Organizations: A Study of SAARC," *Indian Journal of Political Science* 50, no. 1 (January–March 1989): 18–27.

33. Margaret R. Biswas, "Environment and Development in South Asia," *Contemporary South Asia* 1, no. 2 (1992): 181–91.

34. *Indian Express,* May 12, 1997.

35. *Times of India,* May 6, 1997.

36. Muchkund Dubey, "Off to a Slow Start: Regionalism in South Asia," *Times of India,* April 22, 1999.

37. Ananya Mukherjee Reed, "Regionalization in South Asia: Theory and Praxis," *Pacific Affairs* 70, no. 2 (Summer 1997): 235–51.

38. Navnita Chadha Behera, Paul M. Evans, and Gowher Rizvi, *Beyond Boundaries: A Report on the State of Non-official Dialogues on Peace, Security & Cooperation in South Asia* (Ontario: University of Toronto–York University Joint Center for Asia Pacific Studies, 1997).

39. Sanjoy Baru, "South Asian Dialogue: Business of Peace and Security," *Times of India,* June 9, 1999.

40. The Koshi project was a joint India-Nepal dam construction venture on the Koshi tributary of the Ganges in the 1950s.

41. Atiq Rahman, "Sustainable Development and Environment Management: A South Asian Perspective," paper presented at "South Asia at the Crossroads."

42. Mukund Govind Rajan, *Global Environmental Politics: India and the North-South Politics of Global Environmental Issues* (Delhi: Oxford University Press, 1997).

43. Shaukat Hassan, "Environmental Issues and Security in South Asia," Adelphi Paper no. 262 (London: International Institute of Strategic Studies, Autumn 1991).

44. Richard A. Matthew, "Recent Books on Environment, Conflict, and Security," *Journal of Political Ecology: Case Studies in History and Society* 6 (1999), available at ⟨dizzy.library.arizona.edu/ej/jpe/jpeweb.html⟩.

45. Ashok Swain, "Democratic Consolidation: Environmental Movements in India," *Asian Survey* 37, no. 9 (September 1997): 818–32.

46. Ashok Swain, "Flood: An Increasing Menace," *Seminar,* no. 478 (June 1999): 30–33.

47. Since 1992, the Inguri River hydroelectric power complex, consisting of a dam and five hydroelectric power plants, is being jointly managed by two warring groups, the Abkhaz and the Georgians. See Paula Grab and John M. Whiteley, "A Hydroelectric Power Complex on Both Sides of a War: Potential Weapon or Peace Incentives?" in Joachim Blatter and Helen Ingram, eds., *Reflections on Water: New Approaches to Transboundary Conflicts and Cooperation* (Cambridge, Mass.: MIT Press, 2001), 213–37.

48. Björn Hettne, "The Fate of Citizenship in Post-Westphalia," *Citizenship Studies* 4, no. 1 (2000): 35–46.

4

The Promises and Pitfalls of Environmental Peacemaking in the Aral Sea Basin

Erika Weinthal

The Aral Sea Crisis

Shortly before the Soviet Union collapsed, photographs of deserted fishing boats trapped on the exposed seabed of what once was the Aral Sea began to appear in the Western press, revealing to the rest of the world the magnitude of the environmental tragedy confronting the peoples of Central Asia.[1] Within a period of only thirty years—less than the life span of one generation—the population surrounding the Aral Sea has witnessed the drying up of the lake on which it had subsisted for centuries. The immediate cause of the desiccation of the Aral Sea and the collapse of its fishing industry was Moscow's economic policy of favoring cotton monoculture over other economic alternatives. Beginning in the 1960s, Soviet planners increasingly withdrew water from the two main Central Asian rivers feeding the Aral Sea for irrigation to cultivate cotton.[2]

These two main rivers, the Amu Darya and the Syr Darya, originate in the eastern mountains of Central Asia and then flow across the Kara Kum and Kyzyl Kum deserts, respectively, before emptying into the Aral Sea, a large terminal lake in the middle of the desert. Until 1960, about 55 cubic kilometers of water flowed into the sea annually, but by the mid-1980s, the Amu Darya and the Syr Darya barely trickled into the sea. As Soviet planners siphoned off water for irrigation, the shores of the Aral Sea receded by 60–80 kilometers; what once was the fourth-largest lake in the world behind the Caspian, Lake Superior, and Lake Victoria shrank to the sixth-largest and, as of 1988, had bifurcated into two separate bodies of water—a "small" sea in the north and a "large" sea in the south. By 1991 the sea level of the Aral had fallen by about 15 meters, the surface area had

86

been reduced by one-half, and the volume had diminished by two-thirds.[3] Salinity levels have tripled, increasing from 10 grams per liter to over 30.[4]

The withdrawal of water upstream for irrigated agriculture has generated a wide array of environmental, economic, and social problems in the Aral Sea basin. The desiccation of the sea has led to a sharp upsurge in dust storms containing the toxic salt residue from the exposed seabed, and in place of the sea, a new desert began to emerge, referred to as the Akkumy (white sands).[5] Overirrigation resulted in the waterlogging and salinization of soils that support not only cotton but also other agricultural mainstays. The downstream populations are confronting a public health crisis—in part due to the dust and salt storms, but more because of the contamination of drinking water saturated with agricultural runoff containing large amounts of pesticides and herbicides. Compounding the lack of potable water in the Aral Sea delta, poor health conditions, inadequate diet, and frequent childbirth raised the rates of infant mortality to 60 per 1,000 births in Karakalpakstan (an autonomous republic within Uzbekistan) by 1989 and to 75 per 1,000 in Dashhowuz oblast in Turkmenistan by 1988.[6] According to the United Nations, between 1985 and 1990, average infant mortality rates ranged from 36 per 1,000 in Kazakhstan to 58 per 1,000 in Tajikistan and Turkmenistan and have not decreased substantially since the Soviet Union's collapse; between 1995 and 2000, average infant mortality rates ranged from 35 per 1,000 in Kazakhstan to 57 per 1,000 in Tajikistan.[7]

In short, the Central Asian countries are experiencing problems similar to those encountered in many other developing countries, such as extreme poverty, poor health care, economic collapse, and environmental degradation. According to the United Nations (UN) *Human Development Report,* the Central Asian countries are ranked as having "medium human development," with Tajikistan at the low end and Kazakhstan at the high end (see Table 4.1).[8] Whereas during the Soviet period, Moscow provided universal education and health care, the loss of resource transfers from the center has led to a sharp decline in the quality of basic health services and has disrupted the educational system. As one of the first steps toward state building, each of the Central Asian countries changed the language of classroom instruction from Russian to the respective national language, even before they had the necessary textbooks to replace prior Russian ones. Thus, environmental problems precipitated by the Aral Sea crisis are only accentuating the social problems caused by the collapse of the Soviet Union.

This chapter deals with specific efforts to mitigate the Aral Sea crisis by devising a new water-sharing regime in the aftermath of the breakup of the

Table 4.1

Human Development Statistics for Central Asia

Country	HDI rank[1]	Life expectancy at birth (years) 1998	Adult literacy rate (%) 1998	GDP per capita (PPP, U.S.$)[1] 1998
Kazakhstan	73	67.9	99	4,378
Kyrgyzstan	98	68.0	97	2,317
Turkmenistan	100	65.7	98	2,550
Uzbekistan	106	67.8	88	2,053
Tajikistan	110	67.5	99	1,041

Source: United Nations Development Programme (UNDP), *Human Development Report 2000* (New York: Oxford University Press, 2000).
[1]For an explanation of the Human Development Index (HDI) and purchasing power parity (PPP), see the notes to Table 2.1 on page 26.

Soviet Union. In order to assess attempts at fostering environmental cooperation and their implications for regional peace, the chapter proceeds as follows. First, it describes some of the real and potential environmental and nonenvironmental conflicts that existed during the era of the Soviet Union and after its collapse. Second, the chapter illuminates both internally and externally driven undertakings to encourage cooperation in the absence of Moscow. Third, it elucidates the scope of cooperation in the Aral Sea basin. Fourth, the chapter evaluates the usefulness of concepts alluded to in Chapter 1, including "changing the strategic context" and "enhancing post-Westphalian governance," for explaining the form and scope of cooperation in the Aral Sea basin. Although the Central Asian states have strived to ameliorate the environmental effects of the Aral Sea crisis, many of the donor-sponsored efforts have not adequately dealt with the sources of the desiccation of the Aral Sea despite having prevented acute conflict. Indeed, the role of international actors in facilitating environmental peacemaking has been ambiguous, since many of their efforts have neither addressed the direct causes of the Aral Sea's crisis nor reached those directly affected by its loss. Finally, the chapter concludes by suggesting a way to strengthen both environmental cooperation and regional peace in Central Asia by linking water-based cooperation to other activities, particularly in the energy and agricultural sectors.

The Principal Axes of Conflict before the Soviet Collapse

Soviet Premier Mikhail Gorbachev's combined policies of glasnost and perestroika in the mid-to-late 1980s opened new opportunities for individuals to voice their demands and grievances concerning the desiccation of the Aral Sea without fear of reprisal from central and regional government officials. Following the loosening of the political reins, many Central Asian writers and intellectuals began to express their resentment regarding the cotton monoculture in Central Asian society, which they saw as the immediate cause of the Aral Sea tragedy. The opening of the political realm led to the emergence of a loosely organized opposition that sought recourse from Moscow to mitigate the effects of the desiccation of the Aral Sea. One of the first grassroots movements to form in Central Asia was the Committee for Saving the Aral Sea, led by the writer Pirmat Shermukhamedov. At the same time, nascent nationalist movements, such as Birlik (Unity) and Erk (Freedom), also pressed the issue of Aral Sea desiccation in their struggle against Russian dominance. They considered cotton monoculture to be the manifestation of Soviet exploitation and the lack of control Central Asians had over their own destiny.[9] Central Asians were demanding an end to cotton monoculture and the opportunity to direct their own path of economic development—one in which they were not merely producers of raw materials. These nationalist movements used the Aral Sea tragedy to press Moscow for more cultural and religious autonomy along with the elevation of the national languages within the respective Central Asian republics. The rise of these environmental and nationalist movements was part of a growing trend in the Soviet Union in which the Central Asian republics were calling for more regional autonomy while the Baltic republics were pushing for independence.[10]

Yet, as conflict between the Central Asian republics and Moscow over the proper economic policy for Central Asia became more pronounced, these national sentiments also began to have local effects that were not so peaceful. In particular, small-scale ethnic conflicts increased in the late 1980s as nationalist tendencies collided with the increasing scarcity of water and land resources in Central Asia. One of the first signs of festering ethnic tension took place in 1989, when Tajiks and Uzbeks quarreled over land and water rights in the Vakhsh River valley. According to local press accounts, waterlogged soils, resulting from poor irrigation networks and drainage systems, forced the resettlement of a group of Tajiks to another sector of the same kolkhoz, which then required that land be redistributed between the Tajik and Uzbek populations. The Uzbek inhabitants, whose personal plots would be lost, vehemently opposed the transfer of agricultural land.[11] Then, in July 1989, a long-standing dispute between Tajiks and Kyrgyz came to a head over land and water rights in the Isfara-Batken district along the border between the Tajik and Kyrgyz republics. Plans to build an inter-republic canal sparked the dispute when the Kyrgyz perceived that they would not receive sufficient water from the Tajiks to irrigate all of their fields. Finally, in June 1990, violent conflict broke out between Kyrgyz and Uzbeks in the city of Osh, Kyrgyzstan due to the reassignment of Uzbek land for Kyrgyz residential housing. Approximately two hundred deaths resulted.

The Principal Axes of Potential Conflict after the Soviet Collapse

When the Central Asian states achieved independence in 1991, the small-scale conflicts discussed above suggested that the potential fault lines would lie at the intersection of ethnic tensions and environmental scarcities. Yet, with independence, other potential intra- and interstate conflicts also came to the fore, as Central Asian leaders were forced to deal with stagnating economies, collapsing social welfare systems, high levels of corruption, disgruntled populations, an increase in Islamic "fundamentalism," and growing political opposition. Many of these new conflicts were tied to the dynamics of the transition period, during which the Central Asian states were forced to undergo simultaneous political, economic, and national-identity transitions as part of the state-building process. The nature of these multiple transitions in weakly institutionalized states created a domestic political context marked by uncertainty and short time horizons.

One of the early threats to political stability and regional peace in Central Asia was the contestability of the new states' inherited territorial borders. During the Soviet period, borders were constructed in order to divide and rule what was previously known as "Turkestan." Although the Central Asian governments are determined to maintain their current territorial borders, these former republican borders are a particularly sensitive issue, as they do not necessary coincide with homogeneous ethnic populations. For example, Tajikistan might one day contest its borders with Uzbekistan, since Soviet elites placed the legendary and culturally rich cities of Samarqand and Bukhoro—two traditionally ethnically Tajik cities—within Uzbekistan. Moreover, Soviet elites divided the Fergana Valley among the republics of Kyrgyzstan, Tajikistan and Uzbekistan, making the valley the most ethnically intermingled region within Central Asia. The way borders weave in and out among ethnic groupings in these three states has compromised the Fergana Valley and has led analysts to claim that it might be a potential hotbed for ethnic conflict.[12]

Other threats to regional peace emanate from internal state-building processes in which the Central Asian states pledged, to various degrees, to conduct free and fair elections and to undertake market reforms. Kyrgyzstan was the first Central Asian country to open its economy, but many of its attempts at structural adjustment have resulted in the near collapse of the economy. In response to growing domestic discontent, the Kyrgyzstan government chose to slow down the reform process in order to avert further destabilizing an already tenuous political situation. Having watched Kyrgyzstan's economy collapse, the leaderships in Turkmenistan and Uzbekistan refused to embark on economic reform, for they feared a further deterioration of their populations' living standards that could in turn bolster growing political opposition. As economic conditions continue to decline and few efforts at political reform are made, the Central Asian leaders' hold on power has become more uncertain. As the Central Asian states have resisted overtures aimed at democratizing their political systems and creating a viable space for an independent opposition, Central Asian leaders have resorted to exiling and arresting any person who is deemed a threat to their hold on power.

The geopolitics of the broader Central Asian region has created other axes for potential conflict. Here, the combined effects of the relentless conflict in Afghanistan, the civil war in Tajikistan (1992–97), and the rise of Islamic "fundamentalism" in Central Asia have impeded attempts at fostering regional peace. Central Asian leaders such as President Islam Karimov in Uzbekistan and President Askar Akaev in Kyrgyzstan, for instance, are especially fearful that political Islamic movements might spread from Tajikistan and Afghanistan into their respective territories, particularly in the Fergana Valley. Political turmoil in Afghanistan and Tajikistan has enabled the trade in narcotics and weapons to flourish. In 1999, Uzbekistan was rocked severely by political instability. During February of that year, several bombs exploded in Tashkent, the capital of Uzbekistan, in an attempt to assassinate Karimov. During the summer and fall of 1999, a hostage crisis erupted in southern Kyrgyzstan in which ethnic Uzbek hostage-takers opposed to Karimov were demanding the release of their Muslim colleagues imprisoned in Uzbekistan. These outbreaks highlight some of the obstacles Central Asian leaders must overcome in order to promote regional peace. Yet, these concerns have not prompted the Central Asian states to coordinate their policies to address these issues; Uzbekistan, for example, has taken unilateral action to seal its borders, leaving families divided by the Uzbek-Kyrgyz border and thwarting cross-border trade.[13]

Axes of Potential Environmental Conflict

The region's new territorial delineations have provided the foundation on which potential interstate water conflicts could ensue. Following independence, Kyrgyzstan became the upstream riparian in the Syr Darya basin. Uzbekistan and Tajikistan share the middle course and Kazakhstan (specifically, its Shymkent and Qyzylorda provinces) is the downstream riparian. In the Amu Darya basin, Tajikistan is the upstream riparian and Uzbekistan is both a midstream and downstream riparian; its Autonomous Republic of Karakalpakstan and Khorezm province are downstream to the upstream users. Dashhowuz province in Turkmenistan is also a downstream riparian in the Amu Darya basin. The Zarafshon, a smaller drainage basin within the Aral Sea basin, also has been internationalized. The Zarafshon River originates in Tajikistan and flows into Uzbekistan, where almost all of its waters are used in the Bukhoro and Samarqand provinces before it vanishes in the desert, never reaching the Amu Darya.

In Central Asia, these interstate territorial boundaries broke down the Soviet system of interdependence in which the source of a republic's water supply was politically irrelevant. With independence, 98 percent of Turkmenistan's water supply and 91 percent of Uzbekistan's originate outside their borders.[14] Even within Uzbekistan, provinces such as Andijon in the Fergana Valley are left without any indigenous water supplies; all of Andijon's water comes not only from outside the local area, but from outside the country.[15] Thus, Central Asia could be susceptible to interstate conflict, given that the water-poor states Uzbekistan and Turkmenistan are also the ones most dependent on the water resources originating from outside their borders for irrigated crop production. According to the UN Food and Agriculture Organization, between 1991 and 1993, 91 percent of Turkmenistan's cropland and 92 percent of Uzbekistan's were irrigated.[16]

Following the breakup of the Soviet Union, cooperation seemed unlikely due to the complex territorial dimension of the Central Asian water situation. Upstream interests collide with those of downstream users over issues of water quantity and quality. In such instances, the benefits of cooperation are highly asymmetrical and unevenly distributed among the water users. Because each main river of the Aral basin flows across at least three states and in some instances meanders in and out of states many times, it is difficult to prevent the various users from over-appropriating the water resources. The use of the rivers of the Aral Sea basin also involves the generation and distribution of externalities: upstream use of the water for irrigation limits the quantity and affects the quality of the water for downstream users.

Aggravating these asymmetries is the fact that the newly independent Central Asian states had to address the environmental legacy of the Aral Sea tragedy, as well as mitigate the rise of the above-mentioned small-scale conflicts. Simply put, the main challenge for the Central Asian region after independence has been to find the means, in the absence of a hegemonic power, with which to respond collectively to the desiccation of the Aral Sea and prevent potential water conflicts from erupting. The remainder of this section reviews some of these potential water conflicts in the Aral Sea basin and highlights the main obstacles to achieving both environmental cooperation and regional peace.

Upstream-Downstream Disputes in the Syr Darya Basin

In the Syr Darya basin, there are at least two related areas wherein conflicts could arise at the interstate level. The more notable one concerns the different scenarios for managing water releases from the Toktogul reservoir in Kyrgyzstan. Kyrgyzstan controls most of the Naryn River, a tributary of the Syr Darya on which some of the region's main hydroelectric stations, dams, and reservoirs are located. The Toktogul reservoir is the only one with substantial storage capacity; essentially, it determines how much water is released to the lower reservoirs along the Naryn cascade. Soviet planners constructed the reservoir to meet irrigation demands downstream rather than to produce energy. Yet, due to diminishing energy supplies from Russia and the other Central Asian states in the immediate years after independence, Kyrgyzstan has chosen intermittently to operate Toktogul for electricity generation rather than for an irrigation regime. When Kyrgyzstan runs the Toktogul power plant in the winter, the water released must be diverted to a local depression, the Arnasai lowland, because of the winter freezing of the lower Syr Darya; as a consequence, the water does not reach the Aral Sea.[17] If it is a dry year, Kyrgyzstan can reduce the water flow to Uzbekistan in the spring and summer, when the demand for irrigation is at its peak downstream.

As the Toktogul situation illustrates, upstream and downstream states have competing interests and differing capabilities. Nor is this the only potential upstream-downstream conflict in the Syr Darya basin. Upstream use affects not only the quantity of water delivered downstream, but also the quality of the water. Hence, downstream users face a different set of constraints. This is particularly true for the midstream agricultural users in the Fergana Valley (primarily Uzbekistan) and the Golodnaya Steppe (Uzbekistan), as well as for the downstream users in Shymkent (Kazakhstan) and Qyzylorda (Kazakhstan). On the one hand, both midstream Uzbekistan and downstream Kazakhstan want to ensure ample water supplies from Kyrgyzstan. On the other hand, the midstream users' interests deviate from those further downstream. The midstream users need water suitable for agriculture, whereas the downstream users in Kazakhstan are concerned about preventing further shrinking of the sea as well as procuring clean drinking water. They receive water laden with agricultural runoff, which has contributed to the health crisis in Qyzylorda province.

Upstream-Downstream Disputes in the Amu Darya Basin

The Toktogul reservoir is only one example of how the Soviet Union's collapse has politicized control and use of water. Similar conflicts could transpire in the Amu Darya basin between the upstream and downstream states, although the civil war in Tajikistan has so far impeded such conflicts from occurring.[18] Of much greater concern in the upstream reaches of the Amu Darya basin is the issue of dam safety. Many of the river's dams are in need of serious repair and several have already collapsed, leading to flooding and unregulated water releases.[19]

More important in the Amu Darya basin, the demarcation of new inter-state borders has elevated the status of various domestic waterways such as the Kara Kum Canal to the realm of international politics.[20] Currently, Uzbekistan and Turkmenistan divide the Amu Darya water equally at Termez. But this allocation could become an international point of contention, as Turkmenistan wants to keep extending the canal so that it can bring new land under cultivation. Herein lies another potential dispute between two midstream states: both have a clear interest in procuring water for the production of cotton, which continues to provide the bulk of their primary revenue, just as it did during the Soviet period.

The clash between upstream and midstream agricultural interests and the downstream interest in clean, potable water represents a situation in which the disincentives for cooperation are starkest. The downstream users in Karakalpakstan possess limited, if any, bargaining leverage over the upstream users, since they lack resources needed by the upstream users. As a consequence, the downstream users have no choice but to use the contaminated water, filled with waste and effluents, for drinking purposes.

The Fergana Valley

The situation in the Fergana Valley captures the way in which local water issues are intertwined with the broader issues of ethnic identity, economic development, and state formation. Within an area about 300 kilometers long and 20–70 kilometers wide, the Fergana Valley has the highest population density in Central Asia. In Andijon province in Uzbekistan, for example, the population density is over five hundred people per square kilometer.[21] Taking into account that the population is growing at a rate of roughly 2 percent annually, this region could re-experience acute ethnic conflict as individuals are forced to compete for already overstressed and scarce resources.

The Fergana Valley is the backbone for agriculture in Central Asia. Fully 45 percent of the irrigation area of the Syr Darya basin, for example, is located within the Fergana Valley. It contains some of the most vital and productive irrigated areas—such as Jalal-Abad and Osh in Kyrgyzstan; Andijon, Namangan, and Fergana in Uzbekistan; and Khujand (Leninabad) in Tajikistan—all of which rely on the Syr Darya and its tributaries for irrigation. Soviet planners built an extensive network of canals for irrigation to support the production of cotton in the valley. With independence, many of these canals transcend the new political jurisdictions, creating a situation in which water users are in competition with each other. In many cases, different ethnic groups are competing for use of scarce water resources in the Fergana Valley, which could affect interstate relations in the post-Soviet context. Consider the above-mentioned conflict between Tajiks and Kyrgyz over irrigation water in 1989. At the time, the conflict was internal and localized, but after independence such micro-level conflicts have international ramifications arising from the importance of new national identities associated with the formation of statehood. Since micro-level conflicts can reinforce both micro and macro identities, subnational conflicts can turn into interstate conflicts—for example, between Tajiks and Kyrgyz along the new international border in Isfara in Tajikistan and Batken in Kyrgyzstan, which previously were neighboring districts of the same country under Soviet rule.

Internally Driven Cooperation (Early 1992)

When the Central Asian republics unexpectedly gained their full independence in December 1991, it was unclear whether they would continue to share the water resources of the Aral Sea basin and find a solution to mitigate the Aral Sea tragedy. Yet the removal of an external decision-making authority in Moscow did not preclude cooperation. Instead, the five states' water ministers signed an initial agreement on February 18, 1992, on "Cooperation in the Management, Utilization, and Protection of Water Resources of Interstate Sources," wherein the fresh water resources of the region were defined as "common" and "integral."[22] According to the agreement, the Central Asian states "commit[ed] themselves to refrain from any activities within their respective territories which, entailing a deviation from the agreed water shares or bringing about water pollution, are likely to affect the interests of, and cause damage to the co-basin states."[23] They agreed to undertake joint activities to solve the problems related to the drying up of the sea and to determine yearly sanitary water withdrawals based on the availability of water resources.[24] This agreement set up the Interstate Water Management Coordinating Commission (later referred to as the Interstate Commission for Water Coordination, or ICWC), composed of the five countries' ministers of water management. This commission would meet on a quarterly basis to set water management policy in the region and define and approve water consumption limits (broken down by growing and nongrowing periods) for each of the republics and for the entire region.[25]

Although it is remarkable that the Central Asian states, unlike many other developing countries, were able to rapidly conclude a formal water-sharing agreement, this first stage of international institution-building for water management had less to do with mitigating the desiccation of the Aral Sea than with ensuring that interstate cooperation would be sustained for political reasons in the transition period. This is so for several reasons. First, transitions entail political uncertainty; shared fears of what the future would hold without Moscow drove policy actions. Second, the memory of the Osh riots, coupled with those of the other small-scale resource conflicts mentioned above, continued to loom over the Central Asian leaderships during the initial days of independence. Third, the only other system of water management that had existed in the pre-Soviet period had been superseded, so the Central Asian leadership could not revert to pre-Soviet practices of decentralized management.[26] Most important, during the first year after independence rapid cooperation can best be explained primarily by inertia—an unwillingness to disrupt or to depart from past practices, especially since the leadership was still concerned with bringing in the cotton harvest.[27]

However, this initial water agreement, like the Indus Waters Treaty in South Asia (discussed in Chapter 3), was largely concerned with conflict prevention, not with environmental protection. Indeed, it remained unclear whether cooperation would last or whether conflict would ensue once the states began to develop their own political and economic policies associated with the establishment of empirical sovereignty. As part of the domestic state-building process, each newly independent state could begin to pursue divergent strategies to accommodate the particular needs of its population. Some of the Central Asian states might stress food security, whereas others could give priority to obtaining energy self-sufficiency. A situation might arise in which both an upstream and a downstream state would want to expand irrigated agriculture. For instance, Turkmenistan has been seeking to extend the Kara Kum Canal in order to support the additional reclamation of land for irrigated agriculture. At the same time, the 1992 agreement notes that Tajikistan also expressed its interest in increasing its water allocations for irrigated farming. Likewise, Kyrgyzstan had intentions to harness its hydroelectric potential. Immediately after independence, it was seeking foreign partners to help it complete two unfinished dams on the upper reaches of the Naryn for the sole purpose of generating electricity.

In short, this status quo agreement was neither equitable nor environmentally sustainable over the long term, especially if the Central Asians sought to restore the Aral Sea, at a maximum, or to preserve its current size, at a minimum. The immediate post-independence framework agreement incorporated the water-sharing rules applied during the Soviet period, which were based on crop requirements and quotas and paid scant attention to water quality. Thus, the Central Asian states had to amend the agreement to ensure a minimum flow into the Aral Sea while addressing the broader question of water quality.[28]

Externally Driven Cooperation (Mid-1992 to Early 1997)

For this initial attempt at regional cooperation to withstand the demands of state formation, wherein the individual Central Asian states might have pursued independent policies concerning their freshwater resources, intervention on the part of the international community was necessary to avert any potential interstate conflict.[29] Indeed, international intervention was essential because these newly independent states lacked the institutional and legal bases for international cooperation as well as the methods and knowledge to devise institutional structures for a water basin that had previously been fully controlled by an outside authority. These weakly institutionalized and poor states greatly needed aid to help pay the costs of the multiple political, economic, and social transitions away from state socialism.

Members of the international community were willing to provide assistance to the Central Asian states because they feared that, as the newly independent states began to undertake national development programs in which water demands could differ from previous allocations, "conflicts of interests and water disputes" would appear.[30] In its initial report on the matter, the World Bank concluded, "Despite the [1992] water agreements signed after independence of the Republics, the potential for future water disputes cannot be ignored" given the importance of freshwater to economic development for the region.[31]Anticipating that conflict would emerge, the World Bank was operating under the assumption that it had to intervene early because national interests would arise, making cooperation more difficult once the Soviet *nomenklatura* was replaced.

In response to the growing potential for conflicts of interest, the World Bank in particular and the international community in general sought to help the Central Asian states formalize environmental cooperation in the Aral Sea basin. Since the "greening" of the World Bank in the mid-to-late 1980s and the creation of the Global Environmental Facility in the wake of the 1992 "Earth summit" in Rio de Janeiro, the World Bank's role as a funder of environmental projects has increased. The collapse of the Soviet Union and the subsequent Rio conference created a unique moment in time for the World Bank to take on a massive project that could link economic development, conflict prevention, and environmental protection.[32] Moreover, with the international donor community focused on aiding these new economies in transition, environmental protection became one part of the donor assistance programs to eastern Europe and the Soviet successor states that would enable them to advance economic and political reform. In short, the Aral Sea crisis provided a unique "test case" to link economic and political reforms with environment and conflict issues.

The Central Asian governments requested assistance from the World

Bank immediately after the Soviet Union's collapse to help mitigate the ecological and health problems near the Aral Sea. Before the World Bank would intervene, however, it made aid and its involvement contingent on the Central Asian states' devising a new institutional framework for water sharing beyond the 1992 agreement. As part of the World Bank's standard operating procedures, it required a clear signal from the Central Asian states concerning their commitment to regional cooperation. The water-sharing crisis in Central Asia could have been viewed solely as a technical or developmental problem, but in this case the World Bank also perceived it as a political question. According to Syed Kirmani and Guy Le Moigne, the World Bank mission in 1992 "stressed the need for regional cooperation and strong commitment and concerted efforts of the Republics" in order to address adequately the Aral Sea crisis over the long term.[33] The World Bank made clear that its intention was not to "save" the Aral Sea, but rather to help the Central Asian states develop a joint program to ensure the management of a shared resource.[34] In short, environmental protection was the secondary reason for intervention. The way the World Bank framed the Aral Sea basin problem in the early 1990s was similar to the way it had constructed the Indus River basin problem four decades earlier (see Chapter 3). In both cases, the World Bank's behavior was not customary, as it intervened not in order to support specific projects that would generate a revenue stream, but in order to promote regional development and to prevent any potential water disputes from escalating into acute conflicts.

In order to meet conditions for assistance by demonstrating a commitment to interstate cooperation, the Central Asian states had to supplement the original 1992 water-sharing agreement and develop a collective framework for dealing with the Aral Sea crisis. On March 26, 1993, the Central Asian leaders signed in Qyzylorda the "Agreement on Joint Activities for Addressing the Crisis of the Aral Sea and the Zone around the Sea and for Improving the Environment and Ensuring the Social and Economic Development of the Aral Sea Region." This agreement and the accompanying statutes created two apex organizations to the ICWC: the Interstate Council for Addressing the Aral Sea Crisis (ICAS) and the International Fund for the Aral Sea (IFAS).[35]

According to the 1993 institutional structure for water management in the Aral Sea basin, the ICAS and its executive committee (EC) became the leading water-management organization in Central Asia. The EC is responsible for developing policies and programs for addressing the Aral Sea

crisis. The ICAS is composed of twenty-five high-level representatives
from the five states, who meet twice a year to discuss and decide policies,
programs, and proposals put forth by the EC. The EC's charter equates it to
a sovereign government with full powers to plan and implement programs
approved by the ICAS. The heads of state established the IFAS to finance
the Aral Sea programs with contributions from the five states and other
donors. According to the agreement, each basin state was supposed to
allocate 1 percent of its gross domestic product to the fund. At the time, the
Central Asian leaders nominated President Nursultan Nazarbaev of Kazakhstan as head of this fund.

Having found the Central Asians' efforts sufficient, the World Bank in
conjunction with the UN Environment Programme and the UN Development Programme (UNDP) began to work with the Central Asians during
the spring of 1993 to prepare a program framework for the Aral Sea basin.
The resultant "Proposed Framework of Activities" called for seven thematic programs and nineteen urgent projects.[36] The seven main thematic
programs included (1) regional water-resources-management strategy (including improving the efficiency and operation of the dams and reservoirs);
(2) hydrometerological services; (3) water-quality management; (4) wetland restoration and environmental studies; (5) clean water, sanitation, and
health; (6) integrated land and water management in the upper watersheds;
and (7) automatic controls of the basin management authorities (BVOs).[37]
In addition, there was also a supplementary program on capacity building.
The five heads of state approved this program on January 11, 1994, in
Nukus, Karakalpakstan. As a result of this program, the World Bank began
to focus its efforts on implementation of the planned projects, which was
estimated to cost $470 million.[38]

The Scope of Cooperation

Even though the Central Asians signed a new interstate agreement in 1993,
signaling to the international community their seriousness about finding a
solution to the Aral Sea crisis, numerous internal inconsistencies existed.
For example, international legal experts pointed out that the relationship
between the ICAS and the ICWC remained unclear due to the overlapping
of functions and responsibilities in their statutes.[39] A legal study sponsored by the Water Resources Management and Agricultural Production
in the Central Asian Republics (WARMAP) program of the European

Union's Technical Assistance for the Commonwealth of Independent States (TACIS) program concluded that these institutions "are regulated by separate statutes that are not completely streamlined as to the institutions' respective functions."[40]

In fact, both the 1992 and the 1993 accords should have been considered framework agreements (i.e., agreements limited to establishing basic principles).[41] Although they espoused the terminology of international water law, both agreements lacked the teeth to make them effective. Rather, they merely unlocked the door for further negotiations over how to restructure water management in Central Asia while also preventing regional discord. Following the signing of both the 1992 and 1993 agreements, the international community (primarily the World Bank and the European Union) became involved in organizing meetings and workshops to implement and deepen environmental cooperation surrounding these accords.

During the first phase of implementation, the World Bank realized that its Aral Sea Basin Program was progressing more slowly than anticipated.[42] Several donor recommendations intended to ensure that these agreements were not mere formalities were never carried out. For example, the Central Asian leadership failed to appoint a permanent chairperson for the EC to conclude an intergovernmental agreement recognizing the international status of the interstate organizations, or to clarify the mandates and jurisdictions of the new apex organizations. Moreover, the IFAS never fulfilled its overall objective to act as a channel through which the Central Asian states would fund programs to address the Aral Sea crisis.

Detecting the lack of advancement in the World Bank's Aral Sea Program, the U.S. Agency for International Development (USAID) chose to deviate from the World Bank program and redirect its efforts in a more concentrated and limited manner. USAID sought to redefine the scope of environmental cooperation in the Aral Sea basin by dealing only with the potential and real disputes over water management schemes for the Toktogul reservoir. Rather than assuming that the 1992 water sharing agreement was fixed and exhaustive, USAID acted as a catalyst for new negotiations over water allocations in the Syr Darya basin. In focusing on the development of a more specific agreement, pertaining solely to water releases from Toktogul, it organized seminars and workshops to study competing scenarios for the timing of water releases.[43]

By 1997, representatives from USAID were arguing that the process of institutionalizing environmental cooperation was dragging because the foreign community had devoted most of its attention to the old water *nomenklatura*.[44] USAID concluded that the members of the old water *nomenklatura* lacked a real interest in reforming the system for water management, preferring to restructure the system just enough to ensure that they retained their positions of power while also procuring much-coveted foreign assistance. Due to the intransigence that emerged with time on the part of the local water *nomenklatura,* USAID began to work with a completely new Central Asian regional body, the Interstate Council for Kazakhstan, Kyrgyzstan, and Uzbekistan, to reach an agreement over Toktogul.[45]

In short, building institutions for interstate environmental cooperation has not been a straightforward process in Central Asia. The main success of this first period was the conclusion of a Water Resources Management Strategy in 1996 by the Aral Sea Basin Working Group.[46] In response to the international community's displeasure with the slow pace of institutional change, the heads of state then decided to streamline the current institutions for water management. During their meeting on February 28, 1997, they resolved many of the above-mentioned points of contention. For example, they moved the IFAS to Tashkent and turned it into the main apex organization, while also appointing President Karimov of Uzbekistan as its head. The ICAS was dissolved and its functions and subordinate agencies transferred to the IFAS, which meant that the ICWC became subsumed under the leadership of the IFAS.

In sum, environmental cooperation in the Aral Sea basin has not been completely formalized and it is still too early to assess its effectiveness in mitigating the Aral Sea crisis; many ecological and social ills continue to plague the region. Indeed, the local populations in the areas near the Aral Sea frequently complain that they have not seen any improvements in their lives from these international donor programs. This said, there have been no violent conflicts related to freshwater resources following the collapse of the Soviet Union. This absence is striking given the real potential for conflict between upstream and downstream users and among localized users in the Fergana Valley. The ongoing *process* of building environmental cooperation has helped to ensure some level of regional peace in Central Asia, especially pertaining to the question of competition over scarce resources. Given that this process of building institutions for environmental cooperation is still unfolding, the next few sections suggest that environmental cooperation and its prospects for regional peace are best understood in relation to the political events of state breakup and subsequent state formation.[47]

Changing the Strategic Context

Interstate cooperation is unlikely when one state gains more and can use its gains to threaten another state in future interaction, or where the danger of unanticipated defection by one state cannot be hedged against.[48] The Soviet breakup and the subsequent demarcation of new boundaries introduced asymmetric capabilities and interests in the Aral Sea basin; each individual state could no longer trust that the other states would continue to cooperate over water. With the removal of Moscow as an external enforcer, each new state became especially sensitive to others' relative gains during negotiations over new sharing arrangements. This was the case because any initial agreement not only specifies who benefits at the first round but also determines the distribution of benefits for future rounds. These initial framework agreements could constrain the Central Asian states in later rounds if they choose to amend or revoke any of the agreements.

Moreover, since the Central Asian states were dealing with asymmetries created by the physical properties of a resource and its infrastructure, a greater potential existed for reneging on commitments. In Central Asia, the original Soviet conditions conferred most of the benefits to the downstream states and their cultivation of cotton, but the downstream states were unable to prevent the failure of upstream states to abide by current patterns of water sharing, since the upstream states inherited most of the reservoirs and dams. The downstream states could not be certain that Kyrgyzstan would not defect and run its hydropower stations purely for generating electricity. Such defection fears are intertwined with the overarching problem of time, in this case characterized by short time horizons on the part of the leadership and the uncertainty of the transition.

One way, then, in which outside actors could help facilitate cooperation and offset various asymmetries of power contributing to the uncertainty of the transition and to the short time horizons of the Central Asian leadership was to sponsor development projects. This was especially important for inducing the upstream states, which might have to relinquish their natural upstream advantage, to participate in the institution-building process. If Kyrgyzstan built new dams upstream to generate hydroelectric power or even continued to run Toktogul in the winter, these actions would restrict water flows downstream. By choosing whether or not to fund such projects, the international community could directly affect water-sharing patterns. International law, which prescribes the need for agreement from all affected parties before such a development project can take place, provides a power-ful tool for intergovernmental organizations to refuse the disbursement of funds for projects that would cause harm to another state.

However, in not supporting these upstream projects, the international community has had to compensate Kyrgyzstan in other ways to gain its participation. Since the Central Asian states are poor, the international community was able to offer financial rewards as compensation to Kyrgyz-stan to prevent it from exercising absolute sovereignty over the upper reaches of the watershed. The World Bank remunerated both upstream riparians, Kyrgyzstan and Tajikistan, by including provisions in the Aral Sea Basin Program for specific projects in the upper reaches of the basin. Program Six focused solely on integrated land and water management in the upper watersheds, and Program One included provisions for dam safety.[49] Otherwise, Kyrgyzstan and Tajikistan would have lacked in-centives to participate in the water negotiations, given that they are so far removed geographically from the immediate effects of the Aral Sea crisis. By compensating upstream riparians with financial and material assistance, the international community may be able to shift states' interests and alter the strategic context for negotiating institutions for environmental cooperation.

Finally, if an upstream riparian is less powerful militarily and econom-ically than the downstream riparians, external actors may be able to facili-tate new negotiations over water-sharing institutions. This strategy mirrors the way in which USAID has come to the defense of Kyrgyzstan in the USAID-sponsored negotiations over water releases from Toktogul. Here, an external actor has helped an economically and militarily weaker state bargain with its more powerful downstream partners by trying to show the benefits of mutual trades in water and energy resources among Kazakhstan, Kyrgyzstan, and Uzbekistan.

Enhancing (Post-)Westphalian Governance: Building States First

In order to induce cooperation, external actors have violated the newfound sovereignty of the Central Asian states. This fact violates the conventional wisdom that expects such transitional states to jealously guard their newly acquired sovereignty and limit the role that external actors can play in shaping their internal decision-making policies. Ironically, environmental cooperation in Central Asia has required the development of Westphalian governance before it can have an impact on the enhancement of post-Westphalian governance. Through the deployment of financial and material assistance to key domestic actors, external actors have diffused international norms concerning the role and functions of the nation-state in the international system.

One reason why the Central Asian states welcomed an external role in promoting environmental cooperation rather than resisting it on sovereignty grounds is that the Central Asian states are weakly institutionalized and poor. With independence practically thrust on them, the Central Asian states found themselves cut off from the flow of resource transfers and subsidies from the center. Lacking a stream of patronage from Moscow, the newly independent governments found it infeasible to tackle alone the domestic health and environmental problems precipitated by the Aral Sea crisis.

The dire need to replace the loss of patronage from Moscow and simultaneously find a solution to the Aral Sea crisis provided the impetus for the Central Asian states to invite the international community to help them and to sign new agreements and declarations as symbols of action. Yet, in contrast to accepted notions in which international cooperation undermines state sovereignty, the act of engaging in a process of international cooperation in Central Asia is actually having the reverse effect. Even though the incentives to sign new agreements have come from external actors, the act of signing new agreements helps to strengthen state sovereignty. Simply put, to be a "normal" nation-state in the international system requires the ability to engage in acts of foreign relations with other states.[50]

In addition, the Central Asian states favored environmental cooperation because they anticipated that the process of forging linkages to multilateral and bilateral organizations would help them gain a sense of "stateness." The act of engaging in environmental cooperation after independence provided a signal to the international community and, in particular, to Russia that the Central Asian states were now independent, sovereign nation-states. In order to establish borders between themselves and, more importantly, between themselves and Moscow, the Central Asian states needed to consolidate their internal sovereignty. For example, the Central Asian states needed to build independent armies to protect their borders. Here again, externally induced cooperation helped the Central Asians to further disengage from the Russian sphere of influence. In this case, the international community has bestowed on the Central Asians pseudo-military arrangements as another means to strengthen their domestic sovereignty. Uzbekistan, Kazakhstan, and Kyrgyzstan have been willing to participate in U.S.-led military training programs and create a Central Asian Battalion as part of the North Atlantic Treaty Organization's Partnership for Peace program. These attempts by the international community to create a situation of regional peace through either environmental cooperation or military cooperation are favorably received by the Central Asian leadership, since they contribute to the state-building process.

By pushing for an interstate agreement before committing to an assistance program, the World Bank was able to transfer norms derived from international law. For the most part, intergovernmental organizations such as the World Bank will not give money for technical assistance if an international river-basin agreement is not in place first because the international donor and legal community equates cooperation on the basis of fundamental water-law principles with the establishment of an international basin institution.[51] Since the operational procedures of the World Bank prohibit it from intervening and supporting projects that contradict accepted principles of international water law, it could imbue the new institutional agreements for water cooperation with norms from international law.

At the domestic level, international advisers were also trying to impose new norms for water management, which included establishing agencies that resemble Western water and environmental agencies. All the Central Asian states except Turkmenistan have adopted new water codes and laws in order to establish further degrees of separation from Russia and to demonstrate their empirical sovereignty to the international community.[52] In conjunction with the adoption of these laws, international actors seek to play a substantial role in helping the governments rewrite and standardize these national laws so that they correspond to the international agreements regulating water use and distribution.

The international community is, in fact, seeking to impose similar institutional structures throughout the region. In the environmental arena, moreover, they are able to exert influence by linking aid programs to the new international norm of "sustainable development" that gained momentum following the 1992 Earth Summit. In the water sector, USAID, along with TACIS and UNDP, has been promoting "water user associations" within Central Asia as part of numerous projects to encourage both privatization and sustainable development within the agricultural and water sectors.

The Central Asian states have affirmed their commitment to the principles of international law in public statements and declarations. Yet without a domestic tradition of the rule of law, past patterns of prior use and local customs often continue to underlie practices governing water use and allocation. For example, early interviews with several water administrators in Osh revealed that none of them had been informed of the new interstate water-sharing framework.[53] Moreover, they suggested that they conduct cross-border water exchanges with neighboring regions in Uzbekistan in the way that they always have—by picking up the phone.

Environmental cooperation in the Aral Sea basin, therefore, is really about creating the myth of statehood on an empirical level. The Central Asian states understood that in order to become legitimate members of the international system, they have to subscribe to certain modes of behavior—even if, in practice, these are merely symbolic. Some scholars have described this process of adopting international norms as rituals wherein institutions are being constituted worldwide because of the myth of their usefulness.[54]

Likewise, the role of the international community is also contributing to the myth of nation building. During the Soviet period, most of the main scientific research institutes associated with water were located in the downstream states. For example, the All–Central Asian Institute for Irrigation Research (SANIIRI) was based in Uzbekistan. Because of the way in which the Soviets associated regions with areas of expertise, it is common to hear Central Asians generalize about the respective ethnic groups in the region. Accordingly, "the Uzbeks are water people," whereas "the Kazakhs have always been oil people." Since institutes such as SANIIRI generated most of the information and data relating to the water system and irrigation use, researchers associated with the water sector in the other four basin states seldom question "Uzbek" authority on water issues. Central Asians, furthermore, stress the long history of irrigated agriculture that took place in the oases in the Aral Sea basin, many of which are located in present-day Uzbekistan.[55] As a result, the government of Uzbekistan continues to perpetuate the myth of Uzbeks as water people and cotton farmers as part of creating a new Uzbek nation in the aftermath of the Soviet breakup. It also has not hurt that the Western international organizations have designated Uzbekistan as their base for operations. This has enabled Uzbekistan to play a disproportionately large role in influencing decisions in the water sector and to perpetuate the Uzbekistan government's faith in cotton cultivation and irrigated agriculture.

Despite this myth of usefulness, it appears that many of these new interstate institutions are unable to produce the expected results at the national level. Although the international community has stressed the need to establish a new institutional framework to mitigate the Aral Sea crisis, many locals have expressed their frustration that the rules governing water use and allocation have changed very little. They see the international consultants and old *apparatchiki* profiting from the international aid programs, whereas many of the locals believe that they themselves are capable of devising their own solutions to their own problems.[56] Many complain that the international donors have been captured by vested interests. Indeed, many of those responsible for producing the Aral Sea crisis are those now charged with finding the appropriate solutions and have received the bulk of the international funding to do so.

In short, the newly independent states have learned quickly that in order to attract international assistance and enter the international community of nation-states, they need only adopt the jargon of external actors. They recognize that constituting new institutions is a form of compliance with the demands of the international community. Although consultants and representatives of intergovernmental organizations are encouraging Central Asian states to sign new interstate agreements and trying to shape national water policies, much of this institutional restructuring has not actually mitigated the Aral Sea crisis. Rather, the process of environmental cooperation appears to be more about nation-state building than about environmental protection.

Environmental Cooperation and Global Civil Society

Rather than helping to transform domestic state institutions in the direction of greater transparency and democratic accountability or to build transnational or even local civil societies, engaging in environmental cooperation has provided the Central Asian leadership with an opportunity to co-opt the language of the eco-nationalist movements that were mounting a real challenge to the authority and legitimacy of the new governments. The Central Asian leadership feared the re-emergence of a viable opposition similar to that which existed prior to the Soviet breakup. Thus, on the one hand, the Uzbekistan government carried out a campaign to squash all remnants of the opposition,[57] but on the other, it has sought to co-opt the local nongovernmental organization (NGO) movement while restricting its activities. This was done through the creation of state-sponsored environmental movements that could present the new voice of environmentalism on behalf of the government. For example, the Uzbekistan government created the state-sponsored ECOSAN (International Ecology and Health Foundation) to raise foreign funds for state-controlled programs such as a $3.2 million project with the UN Children's Fund (UNICEF) to provide humanitarian aid to children and mothers in the Autonomous Republic of Karakalpakstan.[58] Established in 1992, it claims to have five million members. However, this organization has not resonated with other local NGOs and instead is seen as an attempt by the Uzbekistan government to dismiss the claims of the opposition that it was ignoring the plight of the Aral Sea.

Overall, many of the Central Asian governments have encouraged the birth of a small NGO community for reasons tied to state building rather than the nurturing of civil society. The creation of a highly censored local NGO community has enabled the Central Asian governments to claim that they are truly engaged in political reform. The Central Asian leadership understands that in order to gain Western recognition, they need to show that they are upholding Western notions of liberal democratic society. In Turkmenistan, the government has thus established its own human rights organization—Demokratii i Prav Cheloveka pri Prezidente Turkmenbashi (Democracy and Human Rights under President Turkmenbashi)—while repressing other indigenous organizations such as the Russkoe Obshchina (Russian Society) that could threaten the government's control over society.[59]

The breadth of issues that local NGOs can address is also limited by the political orientation of the regimes, which have all become more closed and repressive in the decade since independence. For example, one of the greatest restrictions on NGO participation in Kazakhstan is the requirement under the Public Associations Law that NGOs must receive prior approval and register with the government.[60] As it turns out, local NGOs realize that they can exist as long as they focus largely on apolitical issues such as the environment.[61]

Regarding the impact of civil society in the discussions over new interstate water institutions, many local and international NGOs express dismay that they have rarely been consulted.[62] Many of the key decisions are being taken by the Central Asian elites in conjunction with external actors such as the World Bank, resulting in the exclusion of various local Central Asian societal groups from the process of bargaining over the nature of the new interstate institutions for environmental cooperation. This elite control, in turn, has hindered the formation of domestic institutions with greater transparency. During the negotiations, only state organizations such as the water ministries and environmental ministries, along with various design and research institutes, have been included in the workshops and seminars. Therefore, the challenge of establishing a transnational civil society in which local and international NGOs can operate with a high degree of autonomy from the state continues to be hampered by the nature of the region's authoritarian political regimes.[63] However, Central Asian governments have not been able to eliminate completely the role of independent local and international NGOs. In fact, NGOs have assumed an important role at the domestic level, occupying a critical space that governments and intergovernmental organizations often overlook. International NGOs provide local assistance and abet societal groups that are excluded from the bargaining process. They do so on three levels. First, international NGOs advocate the inclusion of the local NGOs in the discussions concerning interstate water-sharing patterns and environmental protection in Central Asia. Second, they have undertaken concrete projects at the local level, focusing on local patterns of water use, health issues, and agricultural practices that directly affect communities living near the Aral Sea. Third, they are attempting to create and reinforce local NGOs as the basis for the emergence of a civil society by establishing ties among them as well as to the international NGO community.

One successful example is the U.S.-based Initiative for Social Action and Renewal in Eurasia. Its Seeds for Democracy Project has dispensed small-scale grants to assist environmental groups with institutional development, administrative support, and ecological projects. A Dutch NGO called NOVIB helped to create the Association of Aral Sea Basin NGOs, which lobbies national and international organizations to reform the principles of water usage in the Aral Sea basin and to increase awareness among the local population on health issues.[64]

Another U.S.-based NGO, the International Committee for the Aral Sea, has been fighting to ensure that many of these local NGOs are represented at the various workshops sponsored by the World Bank and at the donors' conferences. One of this group's main success stories, according to its founder, Bill Davoren, was being the only NGO represented at the two main "Participants' Meetings" of donors that took place in Paris in 1994 and Tashkent in 1997.[65] Since the Central Asian water *nomenklatura* competes with the local NGOs for much of the available international assistance, the IFAS (and previously the ICAS) often sees itself as the "gatekeeper" to the international community. Thus, at the 1997 Participants' Meeting, the IFAS at first chose not to invite any of the local NGOs.

However, with much persistence, the International Committee for the Aral Sea, the Union for Defense of the Aral Sea and Amu Darya, and Médecins Sans Frontières (Doctors without Borders) were able to attend the meeting.

Yet, at the same time, international NGOs and other international organizations must be careful not to mislead Central Asians, many of whom had high expectations that Western organizations and not Central Asian governments would be responsible for solving the Aral Sea crisis. Indeed, over time a sense of donor fatigue has appeared. Many local NGOs are skeptical about meeting with foreigners, believing that most foreigners come with empty pockets only to gawk at the Central Asians' tragedy. One often hears that if each foreigner who came to Central Asia brought along a bucket of water, there would be enough water to resuscitate the Aral Sea.

Promoting Regional Peace in Central Asia: Linking Energy and Water Cooperation

By paying attention only to the issue of water immediately after independence, the international community overlooked opportunities to deepen environmental cooperation and simultaneously foster a situation of regional interdependence and peace. When negotiations focus solely on water sharing, differences between upstream and downstream riparians are reinforced as the gains and losses become very clear.[66] Thus, although cooperation did emerge in Central Asia immediately after independence, the way in which the World Bank constructed the problem as purely an issue of water ossified the stark upstream-downstream divide among the states.

Although violent conflicts over land and water resources have subsided, other forms of discord have appeared that have affected the ability of Central Asian states to share their water resources. These new disagreements involve the exchange of energy resources among the Central Asian states. In order to understand why cooperation solely over water has not had a broader impact on regional peace for Central Asia, this section focuses on the intricate relationship between water and energy.[67] The World Bank failed to recognize this relationship in its construction of the water problem. However, once USAID began to link water and energy, new possibilities for fostering regional peace began to emerge.

The Soviet legacy of regional economic specialization sheds light on the unique relationship of interdependence between water and energy. The Soviet division of economic labor meant that some republics specialized in supplying energy in certain forms to other republics, while other energy sources remained untapped. Kazakhstan has a vast supply of unexploited oil reserves but depends on Uzbekistan for natural gas; Kyrgyzstan has the potential to develop hydroelectric power yet is dependent on Kazakhstan and Uzbekistan for oil and gas, respectively; and Uzbekistan has consistently been a large supplier of natural gas to its neighbors but does not control the headwaters of any of the rivers that traverse its territory.

The breakup of the Soviet Union forced the Central Asian states to reconsider previous patterns of energy use and distribution in order to carry out policies that would forge their independence from one another. The Central Asian states confronted an unusual dilemma of mutual energy dependency similar to the Soviet legacy of integrated water management. If one country failed to supply another with the energy it needed, not only economic hardship but also international tension would ensue. As such, the Central Asian states were forced to devise regional institutions for energy cooperation while also creating regional institutions for the management of their shared water resources.

Ironically, the Soviet legacy of economic interdependence could have served as an asset in fostering environmental cooperation and regional peace. Even while external actors such as the World Bank were concentrating their efforts on sector-by-sector analysis, the Central Asian states were still upholding many of these interdependencies. For example, in 1995 and 1996 Kazakhstan, Kyrgyzstan, and Uzbekistan renegotiated their yearly barter agreements between fuel and water resources. According to this barter arrangement, Kyrgyzstan supplies both Uzbekistan and Kazakhstan with water during the summer months in return for gas and coal, respectively, during the winter months.[68]

Yet, as the downstream states began to pursue independent policies in the energy sector, attempts at water cooperation were undermined. Uzbekistan began to charge Kyrgyzstan for its gas deliveries; Kazakhstan privatized its coal industry. Even if Kyrgyzstan abided by the irrigation scheme for Toktogul and delivered water to Uzbekistan and Kazakhstan, it had no guarantee that it would receive gas and coal in return, since it was now expected to pay market prices for these commodities. Thus, Uzbekistan and Kazakhstan could easily cheat, since these framework agreements dealt with water but not with energy allocations. Enforcement mechanisms were absent, and, for that matter, it was not even clear whether the Uzbekistan government would deliver water to Kazakhstan in accordance with the established allocations. In short, the reciprocal relationships that had existed prior to the Soviet Union's breakup had been replaced by a situation in which each newly independent state's decisions were guided by short time horizons during which each state sought to procure as much foreign revenue as possible, as quickly as possible.

As mentioned earlier in this chapter, USAID began to take into account the possibilities of linking energy and water when it sponsored a 1996 roundtable meeting to develop a multiyear agreement for the operation of the Toktogul reservoir. The objective of this and subsequent meetings was to demonstrate to the Central Asians the political and monetary benefits obtained from exchanging fuel and water resources. USAID's efforts resulted in an interstate water compact between Kazakhstan, Kyrgyzstan, Tajikistan, and Uzbekistan, signed by their prime ministers in 1998, dealing with the timing of releases from the Toktogul reservoir. This alternative construction of the Aral Sea basin crisis as a water-and-energy problem suggests that if international actors had recognized these linkages early on, festering conflicts of interests and capabilities over Toktogul might have been thwarted. Challenges still remain regarding how to implement this interstate water compact, however, and regarding how to develop a similar solution for the Amu Darya basin.[69]

Conclusion: What Is to Be Done?

Without international intervention, it is not clear that the Central Asian states would have created new interstate institutions for the management and protection of the waters of the Aral Sea basin and for the Toktogul reservoir. Although the Central Asian states have not engaged in violent conflict over their water resources, it remains questionable whether these efforts are sufficient to mitigate the Aral Sea tragedy. Many Central Asians argue that donor aid has not solved the ecological ills in the region. What may be needed is a more comprehensive program along the lines of the Caspian Environmental Program (discussed in Chapter 6), in which sustainable development is emphasized over a narrower concern for conflict prevention. At the same time, more weight should be given to local initiatives and community participation. One local initiative in Aralsk, Kazakhstan, is leading the way. In 1997, local residents began to build a dike between the two lakes that have resulted from the bifurcation of the Aral Sea.[70] Such efforts have led to the revival of the smaller, northern Aral Sea.

At the same time and more imminently, the Central Asian leaderships must also deal with the role of agriculture in the Central Asian economies if they want to attack the root cause of the desiccation of the Aral Sea. For environmental cooperation to bring about a deeper regional peace, a more comprehensive, multisectoral approach is essential. In order to deal with the ecological ills generated by the Aral Sea crisis, the international community must encourage reform in the agricultural sector, with less water-intensive crops replacing cotton. Future international assistance should aim to help the Central Asians enact such reforms by providing assistance to farms where the soil can no longer support cotton cultivation.

In the long term, bringing agriculture into the equation would be the most efficient strategy to rectify seventy years of disregard for the environment. However, in the short term, linking water and energy with agriculture might overconstruct the water problem in Central Asia, for reasons related to domestic politics. Cotton during the Soviet period served as a mechanism for social control; thus, in the short term, the transition away from cotton monoculture could threaten governments' means of social control and stability—particularly in Turkmenistan and Uzbekistan, which both rely heavily on the sale of cotton for foreign exchange.[71] However, without addressing the agricultural issue, the chance that environmental cooperation will strengthen regional peace will decrease.

One lesson from the Aral Sea is that international actors can shape the bargaining dynamics over resource issues. They have power, through the financial and material resources at their disposal, to construct the bargaining game in a region, with a direct impact on the likelihood of regional peace. For Central Asia, a strategy premised on broader issue-linkages might still help to strengthen environmental peacemaking. Linking sectoral negotiations over resources such as energy and water may provide mutual gains to both upstream and downstream states, reducing uncertainty concerning the fear of defection. Instead of trying to break down the interdependencies of the Soviet system as part of the state-building process, Central Asian governments and the international community should seek to build on this legacy of interdependence to realize mutual benefits.

At the same time, the Aral Sea basin provides another lesson: linking environmental issues and peacemaking might not promote environmental sustainability unless local actors are involved. Although those who were responsible for draining the Aral Sea are now responsible for restoring it, those who have borne the brunt of the environmental damage continue to be left out of many of the large donor projects. Thus, international organizations need to build linkages between large programs designed to promote regional cooperation and small programs designed to improve the livelihoods of the people in the Aral region. Whereas most large-scale donor efforts have failed to meet their objectives in the region, local NGOs have been initiating several successful programs outside of these regional programs. For example, Perzent, an NGO established in 1992, has carried out numerous programs to improve the status and health of women and children in the Aral Sea region of the Autonomous Republic of Karakalpakstan.[72] Given that many local and international NGOs are addressing many of the immediate health and environmental issues in the Aral region, their initiatives should be incorporated into the larger regional programs. This would require opening up the policymaking sphere to new actors who are not part of the former water *nomenklatura* or the cotton lobby.

Notes

1. For example, see William Ellis, "The Aral: A Soviet Sea Lies Dying," *National Geographic,* February 1990, 73–92.

2. By the mid-1980s there were approximately 7.2 million hectares of irrigated land in Central Asia; in contrast, in 1950 there had been only 2.9 million hectares. See Michael H. Glantz et al., "Tragedy in the Aral Sea Basin: Looking Back to Plan Ahead?" in Hafeez Malik, ed., *Central Asia: Its Strategic Importance and Future Prospects* (New York: St. Martin's, 1994), 167–68.

3. Philip Micklin, "The Aral Crisis: Introduction to the Special Issue," *Post-Soviet Geography* 33, no. 5 (1992): 275.

4. Micklin, "The Aral Crisis," 274–75.

5. David R. Smith, "Change and Variability in Climate and Ecosystem Decline in Aral Sea Basin Deltas," *Post-Soviet Geography* 35, no. 3 (1994): 142–65.

6. Philip P. Micklin, "Water Management in Soviet Central Asia: Problems and Prospects," in John Massey Stewart, ed., *The Soviet Environment: Problems, Policies, and Politics* (Cambridge, U.K.: Cambridge University Press, 1992), 103.

7. See the website of the UN Economic and Social Commission for Asia and the Pacific (ESCAP): ⟨www.unescap.org⟩.

8. UN Development Programme (UNDP), *Human Development Report 2000* (New York: Oxford University Press, 2000).

9. James Critchlow, *Nationalism in Uzbekistan* (Boulder: Westview, 1991), xii.

10. For example, see Jane Dawson, *Eco-Nationalism* (Durham: Duke University Press, 1996).

11. Foreign Broadcast Information Service (FBIS), *Daily Report: Central Eurasia,* June 16, 1989 (FBIS-SOV-89-115), 49.

12. Nancy Lubin, Keith Martin, and Barnett R. Rubin, *Calming the Ferghana Valley: Development and Dialogue in the Heart of Central Asia* (New York: Council on Foreign Relations Press and Century Foundation, 2000); and Anara Tabyshalieva, *The Challenge of Regional Cooperation in Central Asia: Preventing Ethnic Conflict in the Fergana Valley* (Washington, D.C.: United States Institute of Peace Press, 1999).

13. Nick Megoran, "Bad Neighbors, Bad Fences," Radio Free Europe/Radio Liberty, March 13, 2000.

14. David Smith, "Environmental Security and Shared Water Resources in Post-Soviet Central Asia," *Post-Soviet Geography* 36, no. 6 (1995): 361.

15. Ibid., 359–61.

16. World Resources Institute, *World Resources 1996–97* (New York: Oxford University Press, 1996).

17. For details see, D. F. Solodennikof, "Issues of the Management of the Toktogul Reservoir under Current Conditions of a Joint Use of the Syr-Darya Basin Water Resources by Kyrgyzstan, Uzbekistan, Tajikistan, and Kazakhstan," *Aral Herald (Central Asian Scientific Tribune)* no. 1 (Spring 1996): 17–22.

18. Afghanistan is also an upstream riparian in the Amu Darya basin, and for any long-term political solution to the water-sharing situation in Central Asia, it will need to be included in the negotiation process. The absence of a clear political authority in Afghanistan has precluded its participation in the negotiations, and as with Tajikistan, decades of political upheaval have prevented Afghanistan from taking steps to alter the water flows along the Amu Darya.

19. One of the main dams that has attracted international concern is Sarez Lake Dam. See "Tajik Lake Might Cause Disaster," Associated Press, January 19, 1998.

20. The Kara Kum Canal is the longest canal in the world, extending for more than 1,300 kilometers. It initially diverts water from the Amu Darya at Kerki near the Afghan border and then transports it across the desert though Ashgabat, the capital of Turkmenistan.

21. John Schoerberlein, "Between Two Worlds: Obstacles to Development and Prosperity," *Harvard International Review* 22, no. 1 (Winter/Spring 2000): 56–61.

22. "Agreement Between the Republic of Kazakhstan, the Republic of Kyrgyzstan, the Republic of Uzbekistan, the Republic of Tajikistan, and Turkmenistan on Cooperation in Management, Utilization, and Protection of Water Resources of Interstate Sources" (mimeograph), Article 1.

23. Ibid., Article 3. See also Technical Assistance for the Commonwealth of Independent States (TACIS), *Water Resources Management and Agricultural Production in the Central Asian Republics,* Vol. 6: *Legal and Institutional Aspects* (Tashkent: TACIS, January, 1996), 8.

24. "Agreement on Cooperation in Management, Utilization, and Protection," Article 4.

25. Ibid., Article 7.

26. See ch. 4 of Erika Weinthal, *State Making and Environmental Cooperation: Linking Domestic and International Politics in Central Asia* (Cambridge, Mass.: MIT Press, 2002).

27. World Bank, Europe and Central Asia Region, *The Aral Sea Crisis: Proposed Framework of Activities* (Washington, D.C.: World Bank, March 29, 1993), 19.

28. TACIS, *Legal and Institutional Aspects.*

29. For further details, see Weinthal, *State Making and Environmental Cooperation.*

30. World Bank, *The Aral Sea Crisis,* vi.

31. Ibid., ii.

32. Similarly, on the changing role of donor institutions in eastern Europe and how they set the agenda, see Barbara Connolly, Tamar Gutner, and Hildegard Bedarff, "Organizational Inertia and Environmental Assistance to Eastern Europe," in Robert O. Keohane and Marc A. Levy, eds., *Institutions for Environmental Aid* (Cambridge, Mass.: MIT Press, 1996): 281–324; and Tamar Gutner, "Cleaning Up the Baltic Sea:

The Role of Multilateral Development Banks," in Stacy D. VanDeveer and Geoffrey D. Dabelko, eds., *Protecting Regional Seas: Developing Capacity and Fostering Environmental Cooperation in Europe* (Washington, D.C.: Environmental Change and Security Project, Woodrow Wilson International Center for Scholars, 1999).

33. Syed Kirmani and Guy Le Moigne, *Fostering Riparian Cooperation in International River Basins: The World Bank at Its Best in Development Diplomacy,* World Bank technical paper no. 335 (Washington D.C.: World Bank, 1997), 14.

34. Ibid., 15.

35. For an overview of all the agreements, see TACIS, *Legal and Institutional Aspects;* and Laurence Boisson de Chazournes, "Elements of a Legal Strategy for Managing International Watercourses: The Aral Sea Basin," in Salman M.A. Salman and Laurence Boisson de Chazournes, eds., *International Watercourses: Enhancing Cooperation and Managing Conflict,* World Bank technical paper no. 414 (Washington, D.C.: World Bank, June 1998), 60–71.

36. World Bank, *Aral Sea Basin Program—Phase 1, Progress Report 2* (Washington, D.C.: World Bank, September 1995), 1, notes that two of the projects were later merged, leaving eighteen individual projects.

37. "BVO" is the Russian acronym for the basin-wide agencies for water allocation that were created in the late 1980s to manage the cascade of reservoirs and the water withdrawal facilities and pumping stations.

38. This estimate was for the first three years. See World Bank, *Aral Sea Basin Program—Phase 1, Progress Report 2,* iii.

39. TACIS, *Legal and Institutional Aspects,* 1.

40. Ibid. For further clarification, see Dante A. Caponera, "Legal and Institutional Framework for the Management of the Aral Sea Basin Water Resources," Tashkent, April 1995.

41. TACIS, *Legal and Institutional Aspects,* 9.

42. World Bank, *Aral Sea Basin Program—Phase 1, Progress Report 3* (Washington, D.C.: World Bank, February 1996), i.

43. For details, see Erika Weinthal, "Sins of Omission: Constructing Negotiating Sets in the Aral Sea Basin," *Journal of Environment and Development* 10 (March 2001): 50–79.

44. Interview with Barbara Britton, USAID, Environmental Policy and Technology Project, Almaty, Kazakhstan, March 12, 1997.

45. Later, this body was expanded to include Tajikistan.

46. Each of the five independent Central Asian states produced a report titled "Basic Provisions for the Development of the National Water Management Strategy," which were synthesized into an October 1996 report titled "Fundamental Provisions of Water Management Strategy in the Aral Sea Basin."

47. On the relationship between state formation and environmental cooperation, see Weinthal, *State Making and Environmental Cooperation,* esp. chs. 3 and 8.

48. On the effects of time on cooperation, see John C. Matthews III, "Current Gains and Future Outcomes: When Cumulative Relative Gains Matter," *International Security* 21, no. 1 (Summer 1996): 112–46.

49. For an overview, see World Bank Preparation Mission, *Aral Sea Program—Phase 1, Aide Memoire,* Vol. 2 (Washington, D.C.: World Bank, March 1994).

50. Connie L. McNeely, *Constructing the Nation-State: International Organization and Prescriptive Action* (Westport: Greenwood, 1995). See also Abram Chayes and

Antonia Handler Chayes, *The New Sovereignty: Compliance with International Regulatory Agreements* (Cambridge, Mass.: Harvard University Press, 1998).

51. For a discussion of international water law, see Stephen C. McCaffrey, "Water, Politics, and International Law," in Peter H. Gleick, ed., *Water in Crisis: A Guide to the World's Fresh Water Resources* (New York: Oxford University Press, 1993).

52. Dates for the adoption of new water codes and laws are the following: Kazakhstan, March 11, 1993; Kyrgyzstan, January 14, 1994; Tajikistan, December 27, 1993; and Uzbekistan, March 6, 1993.

53. Interview with the head of Osh Oblastvodkhoz (Osh Water Authority), Osh, Kyrgyzstan, April 27, 1995.

54. See John W. Meyer and Brian Rowan, "Institutionalized Organizations: Formal Structure as Myth and Ceremony," in Walter W. Powell and Paul J. DiMaggio, eds., *The New Institutionalism in Organizational Analysis* (Chicago: University of Chicago Press, 1991), 41–62.

55. See, for example, Akmal Karimov, "History of Irrigation in Uzbekistan and Present Problems," April 1997 (photocopy).

56. For example, interview with Vadim Igorevich Antonov, Tashkent, Uzbekistan, January 24, 1995.

57. For an overview of the way that the Uzbekistan government has suppressed any form of dissent, see Abdumannob Polat, "Central Asian Security Forces against Their Dissidents in Exile," in Roald Z. Sagdeev and Susan Eisenhower, eds., *Central Asia: Conflict, Resolution, and Change* (Chevy Chase, Md.: Center for Political and Strategic Studies Press, 1995), 197–216.

58. UNDP Regional Project, "Aral Sea Basin Capacity Development," *Aral Sea Basin NGO Directory* (Tashkent: Aral Sea Basin Capacity Development Project, 1996). In addition, World Bank, *Aral Sea Basin Program—Phase 1, Progress Report 3,* 9, notes that UNICEF is conducting similar-size programs in the disaster zones of Kazakhstan and Turkmenistan.

59. Correspondence with Cassandra Cavanaugh of Human Rights Watch.

60. Scott Horton and Alla Kazakina, "The Legal Regulation of NGOs: Central Asia at a Crossroads," *CIS Law Notes,* March 1998, ⟨www.pbwt.ru/Resources/newsletters/cis98a004.html⟩.

61. Pauline Jones Luong and Erika Weinthal, "The NGO Paradox: Democratic Goals and Non-Democratic Outcomes in Kazakhstan," *Europe-Asia Studies* 51, no. 7 (November 1999): 1267–84.

62. This sentiment was articulated in a formal response by Oleg Tsaruk of the Law and Environment Eurasia Partnership/Aral Sea International Committee to the paper delivered by Philip Micklin at the Social Science Research Council's Water Conference, Tashkent, Uzbekistan, May 19–21, 1998.

63. On global civil society, see Ronnie D. Lipschutz, *Global Civil Society and Global Environmental Governance: The Politics of Nature from Place to Planet* (Albany: State University of New York Press, 1996).

64. Cited in Ecostan News, February 1997 and June 1997: ⟨www.ecostan.org/Ecostan/enindex.html⟩.

65. Correspondence with Bill Davoren, June 1998. In addition, some representatives from the World Bank have informally expressed their reservations about having NGOs participate in meetings and seminars.

66. John Waterbury, "Transboundary Water and the Challenge of International Cooperation in the Middle East," in P. Rogers and P. Lydon, eds., *Water in the Arab World: Perspectives and Prognoses* (Cambridge, Mass.: Harvard University Press, 1994), 39–64.

67. See Weinthal, "Sins of Omission."

68. "Protocol of the Meeting of Representatives of Fuel-Energy and Water Management Complexes of Kazakhstan, Kyrgyzstan, and Uzbekistan on Problem of the Toktogul Cascade Water-Energy Resources Use in 1996," *ICWC Bulletin* 11 (November 1996). According to this arrangement, Uzbekistan and Kazakhstan will also buy the excess energy generated during the summer months by the operation of Toktogul.

69. Correspondence with Daene McKinney, consultant, USAID.

70. Judith Matloff, "Optimism Rises, with Water, in Bid to Revive the Aral Sea," *Christian Science Monitor,* February 5, 1999, ⟨www.csmonitor.com/durable/1999/02/05/p8s2.htm⟩.

71. Weinthal, *State Making and Environmental Cooperation,* ch. 4.

72. See M. Holt Ruffin and Daniel C. Waugh, eds., *Civil Society in Central Asia* (Seattle: University of Washington Press, 1999).

5

Environmental Cooperation for Regional Peace and Security in Southern Africa

Larry A. Swatuk

In 1972, Hedley Bull wrote that "the sources of facile optimism and narrow moralism never dry up, and the lessons of the 'realists' have to be learnt afresh by every new generation. . . ." If the academic study of international relations can find little save period-piece interest in the ideas of the classical realists, that is more a comment upon the competence of scholarship today than upon any change in world conditions.

—Colin Gray[1]

Projects for change and progress need to be linked to the possibilities of society, national and international, as given. Fatalism in the face of the given is as unnecessary as speculation unhinged from practicality and real movement. We need, in [Walter Garrison] Runciman's judicious phrasing most pertinent to [international relations], to distinguish the "improbably possible" from the "probably impossible." This involves the double assertion—one intellectual, the other sociological: the intellectual revolves around a reassertion, chastened by history and critique alike, of the values associated with the Enlightenment; the sociological involves an assertion that, within the constraints of the contemporary world, and of that modernity which characterizes it, purposive action, linked to agency by individuals, movements and states alike, is possible.

—Fred Halliday[2]

In southern Africa today, parallel, opposed narratives are regularly used to explain contemporary events and guide policy decisions. One reflects the classical realist assertions of Gray, the other the Kantian liberalism of Halliday in the epigraphs above.[3] Each provides compelling evidence in support of its analyses. Kantian narratives argue that three dominant themes drive regional developments: the post-apartheid democratic moment; the socio-economics of neoliberal structural adjustment, now including a "regional integration" component; and the post–Rio summit emphasis on the collective management of natural resources. The manner in which these interlink and overlap forms the basis for the emergence of more positive and constructive inter-, intra-, and trans-state relations: a new language fueling new thinking about a new regionalism.[4]

Supporters of the Kantian perspective usually emphasize the functional character of the new regionalism, with the Southern African Development Community (SADC) and transnational, subnational, and multilateral activity at the forefront of analysis.[5] An emerging regional energy grid, an impending trade protocol, communications and transportation networks, transnationally managed "superparks" (peace parks that straddle national boundaries), and new water laws, institutions, and protocols are all said to be moving the region toward cooperation and peace. The post-apartheid democratic moment, therefore, facilitates the collective pursuit of international norms in the region. These values of peace, economic growth, and environmental sustainability depend on increasing space for civil society to articulate its diverse needs and interests and exercise its capabilities free from a domineering and overdeveloped state. They also require the more efficient allocation of often-scarce resources by state-makers in the region and the development of creative and effective partnerships among relevant stakeholders (the state, corporations, nongovernmental organizations [NGOs], and community-based organizations).[6]

To be sure, the Kantian argument continues, there are many obstacles along the road to regional peace and security. Like much of sub-Saharan Africa, the SADC region is characterized by weak, distorted, and divided economies, war-battered societies, and too-often unresponsive governments. Yet, unlike most of Africa, the SADC region enjoys the confidence and concern of the industrialized world. As a result, there seems to be much concentrated global effort toward making the "African renaissance" a reality, at least in this region.

For an increasing number of observers, however, it is the obstacles that mark the proper point of departure in regional analysis.[7] State-makers in Africa's Great Lakes region—which comprises the Democratic Republic of the Congo (DRC), Uganda, Rwanda, and Burundi—and Angola are without a doubt engaged in Machiavellian-Clausewitzian practices of "statecraft." Others—Zimbabwe, Zambia, Namibia, Botswana, and South Africa—look on with varying degrees of interest. Levels of participation are partially determined by the immediacy of geopolitical events and cost-benefit analysis. Personalities and personal rule—the essence of Bismarck and Talleyrand, of classical realism, and so belittled in analyses of African politics over the last three decades—seem once again to be playing decisive roles in policymaking.

Even in those countries where bureaucratic structures of decision making are more firmly entrenched—i.e., Botswana and South Africa—calculations are increasingly made on the basis of narrow "national interests." Ironically, states may be seen to be moving, simultaneously, in two directions: South Africa's intelligence, defense, and foreign policy communities continue to resist participation in the U.S. African Crisis Response Initiative and remain ambivalent about SADC's Organ on Politics, Defense, and Security, but clearly favor bilateral "defense and security" initiatives with members of the North Atlantic Treaty Organization and the European Union.[8] In other words, they are busy building "national fences" in a "dangerous region." At the same time, departments of water affairs, tourism, trade, and industry, among others, are actively seeking ways of tearing and keeping these national fences down.

The environment, quite literally, stands at the center of these contradictions. Whereas the border areas between Namibia, Angola, and Botswana have become sites of conflict, these countries share the Okavango River basin and are party to the international Okavango River Basin Commission (OKACOM). Similarly, the Zambezi River basin marks a zone of cooperation through the large, multilateral Zambezi River Action Plan (ZACPLAN), but that river also forms an area of cross-border conflict among Angola, Zambia, the DRC, and Zimbabwe. Interestingly, in each case, conflict flows with the run of the river: headwater states are deep in conflict; those at mid-flow stand poised between Janus and Minerva; and those at the mouth feel relatively helpless, as they contribute virtually nothing to the flow but accumulate everyone's effluent. States in this last category are the most keen on multilateral agreement. Depending on whether one favors Kant or Machiavelli, or privileges cooperation or conflict, the environment appears either as the locus of opportunity for regional peace-building or as a fragmented set of resources to be defended or captured.

To say that conflict and cooperation are characteristic of the region is stating the obvious. A pertinent question to ask, however, is which of these trends is likely to dominate regional relations during the first decade of the twenty-first century? And, as a corollary, can deliberate emphasis on cooperative tendencies in one issue area help foster abatement of conflict in another? In other words, without losing sight of Gray's remarks, are we justified in acting on Halliday's encouragement?

This chapter examines these questions through an environmental lens. It suggests that overlapping ecological interdependencies do lead toward the development of common resource management regimes; that trends toward post-Westphalian governance can lead toward more sustainable and equitable development practices in the region; and that emphasis on functional, often low-political issue areas can bind regional state-makers and stakeholders together in such a way as to offset the divisive, negative, and potentially conflictual reality of weak formal political economies. To conclude that these developments help move the region toward a peaceful and (shared) prosperous future, however, would be to lose sight of the "improbably possible" and to border on the "probably impossible." Nevertheless, "fatalism in the face of the given" is not only unnecessary, it does a deep injustice to those people actively seeking means to achieve more inclusive and lasting forms of peace and security in southern Africa.

In support of these arguments, the chapter proceeds in the following way. First, it locates the southern African region in terms of historical patterns of human settlement. It then examines the varied development trajectories emerging from the region's history of incorporation into a global economy and summarizes their impacts on the environment. This, then, forms the backdrop for systematic assessment of the region's real and potential sites of conflict, as well as for a discussion of extant patterns of environmental cooperation and conflict. The chapter highlights cases involving Angola, Lesotho, the DRC, Botswana, and Namibia. Lastly, some conclusions are drawn regarding the changing strategic climate in the region, including the role of "post-Westphalian" forms of governance.

Historical Patterns of Human Settlement

Although it is difficult and often misleading to generalize about a region that boasts such a complex human and physical geography, certain trends can nevertheless be discerned. Perhaps most fundamental is the way the historical pattern of human settlement—in particular, the advent of colonialism and the establishment of settler societies—determined the setting for and direction of the environmental problems facing the region today.

The establishment of settler societies in southern Africa mirrors what Alfred Crosby describes as the search for "neo-Europes."[9] Although the violence of European conquest is not unique to the region, the congruence of particular factors—imperialism, social Darwinism, industrialism, and military innovation—ensured that human interventions were particularly upsetting to hitherto sustainable biotic systems.

Nodes of accumulation were established based on the region's insertion into the evolving global capitalist system. The most fertile lands were committed to plantation agriculture in parts of Angola, Zambia, Zimbabwe, Mozambique, Botswana, Swaziland, and South Africa; less fertile areas were devoted to pastoralism in all of the above plus Namibia. In every case, cash crops displaced food crops and indigenous flora and fauna made way for exotics. Ports were developed up and down the Atlantic and Indian Ocean coastlines, where railheads were established; these railway lines often extended no further into the interior than the nearest mine-head. Large urban areas developed around these points of mineral extraction. Primary and secondary industrial production grew up around mineral development and agricultural and pastoral activities.

Several relevant points bear mention here. First, because of mineral development, "neo-European" settlements extended well beyond naturally fertile and easily habitable zones into arid and relatively inhospitable areas—from well-watered coastal plains to the dry interior, for example. Efforts to mimic European lifestyles made human interventions into the regional biota particularly violent (the Orange River basin, for instance, is home to twenty-eight dams and seven water-transfer schemes) and in all likelihood unsustainable in the long term. Second, imperialist competition carved up the region into a series of mutually exclusive geographical, juridical entities, thereby ensuring minimal interlinkages and interactions during the mutually reinforcing periods of (colonial) "incorporation" and (post-colonial) "modernization." Third, as the region became more deeply incorporated and centrally involved in global political economic processes, South Africa increasingly came to play the role of regional staging point. This "sub-imperialist" role[10] is reflected in, among other things, a transportation network geared more than ever to the transfer of raw materials out of southern Africa, and finished goods into the region.[11]

Fourth, settled agriculture and mineral-based development combined to upset the natural "flow" of activities, including the movement of goods and people, across the region. Enclave development was one consequence of modernization at the global periphery: plantations and mines, in particular, drew hundreds of thousands of workers from all over southern Africa. Today, for example, "about 55 percent of the total population [of Zambia] is concentrated in a band about 60 kilometers wide on each side of the railway line from Livingstone to the Copperbelt where the bulk of commercial, industrial, mining and other ancillary economic activities takes place."[12]

Today, "legitimate" migrants to South Africa's mines are being replaced by "illegal" migrants to urban centers. As South Africa's gold mines dwindle and the flow of mine workers is reversed, the "promise" of *Egoli* (gold) has brought vast numbers of people, estimated at several million, into South Africa. In South Africa, roughly 70 percent of the population resides in a cluster of five conurbation areas: greater Cape Town, greater Port Elizabeth, Johannesburg-Pretoria, greater Durban, and greater East London. Each of the rest of the SADC countries is characterized by a dominant city, in each case the capital, where a relatively small upper-and-middle class finds itself increasingly ringed by impoverished squatters.

Fifth, this urban-migration trend leads to a more general observation about regional "flow regimes." Given that regional "development" has, for several hundred years, been oriented toward European tastes and often-competing interests, southern Africa's human and natural resources have been alienated from each other. Natural resources, such as rivers, which used to play the role of life-giving regional arteries, came to be viewed by Europeans as convenient boundaries. Hence, families, clans, and ethnic groups straddling the Cunene, Orange, Zambezi, and Limpopo rivers were separated by the exigencies of Westphalian state-building.[13] Of equal importance were the processes underway within states: race, more than class, became the principal consideration in the organization of space.

Finally, and perhaps most important, local peoples and practices in every case were marginalized from the "development" process, with South Africa's "homelands" and "townships," and Zimbabwe's so-called high-density suburbs being the most visible but hardly the only outcomes of this process. Thus, re-envisioning development as a process by and for southern Africans—and, more fancifully perhaps, as a process that recovers the natural rhythm of the regional ecology—is a necessary part of overcoming historical processes of underdevelopment. Yet, as this chapter demonstrates, there is powerful resistance to this idea.

Patterns of Regional "Development"

The cumulative impact of this history has been overwhelmingly negative on the structure of formal political economies, on peoples' lives, and on the

Table 5.1

Human Development Statistics for the SADC Countries

Country	HDI rank[1]	Life expectancy at birth (years) (1998)	Adult literacy rate (%) (1998)	GDP per capita (PPP, U.S.$)[1] (1998)
"Medium" human development				
Seychelles	53	71	84	10,600
Mauritius	71	71.6	83.8	8,312
South Africa	103	53.2	84.6	8,448
Swaziland	112	60.7	78.3	3,816
Namibia	115	50.1	80.8	5,176
Botswana	122	46.2	75.6	6,103
Lesotho	127	55.2	82.4	1,626
Zimbabwe	130	43.5	87.2	2,669
"Low" human development				
Dem. Rep. of the Congo	152	51.2	58.9	822
Zambia	153	40.5	76.3	719
Tanzania	156	47.9	73.6	480
Angola	160	47.0	42	1,821
Malawi	163	39.5	58.2	523
Mozambique	168	43.8	42.3	782

Source: United Nations Development Programme (UNDP), *Human Development Report 2000* (New York: Oxford University Press, 2000).
[1]For an explanation of the Human Development Index (HDI) and purchasing power parity (PPP), see the notes to Table 2.1 on page 26.

natural environment. In terms of peoples' lives, according to the United Nations Development Programme, eight SADC member states are ranked as having "medium" human development, and six have "low" human development (see Table 5.1).

In terms of formal political economies, the World Bank's *World Development Report* describes SADC's members as being low-income (Mozambique, Tanzania, Malawi, Zambia, Angola, Zimbabwe) or middle-income (with Lesotho, Namibia, and Botswana categorized as "lower-middle" and South Africa and Mauritius as "upper-middle") economies.[14] Although the World Bank's data primarily focus on macro- and microeconomic factors, it also includes tables on "population and labor force," "land use and urbanization," "commercial energy use," and "forest and water resources." And even though average rates of growth in gross domestic product (GDP) for the 1985–95 period were generally positive—particularly in Botswana (7.1 percent), Lesotho (7.0 percent), and Mozambique (5.8 percent)—when adjusted for population growth, SADC's formal economies are more or less stagnant. Moreover, the high growth rates registered in Botswana, Lesotho, and Mozambique have been driven by government-led construction booms (in mineral exploration, the Highlands Water Project, and postwar reconstruction, respectively) and so are both narrowly defined and prone to downward cycles.

The region's and the continent's economic "powerhouse," South Africa, has consistently shown post-apartheid growth rates of around 0 percent when adjusted for population growth. Moreover, there has been an estimated net loss of some 940,000 jobs in South Africa during the post-apartheid era. The region's other relatively diversified and industrialized economy, Zimbabwe's, has been in economic free fall since early 1998 due to a combination of events: President Robert Mugabe's expropriation of white-owned farms, his cash-strapped government's continued willingness to meddle in the DRC's civil war, continuing domestic instability following the 2000 general and 2002 presidential elections, and the "Asian contagion" of financial instability. These factors have prompted a five-year-long run on the Zimbabwean dollar.

SADC economies remain overwhelmingly dependent on the export of one or a few primary products: oil in Angola, diamonds in Botswana, water in Lesotho, tobacco in Malawi, fish in Mozambique, minerals in Namibia and South Africa, sugar in Swaziland, coffee in Tanzania, copper in both the DRC and Zambia, and tobacco and minerals in Zimbabwe. As stated earlier, these violent, extractive processes begun during colonial times reinforce structural inequalities: between city and countryside, white and black, traditional elites (such as the baKgatla tribe in Botswana) and traditionally oppressed (such as the San or "bushmen" in Botswana, Namibia, and South Africa), agro-industry and smallholder agriculture, the built environment and the natural world. Unfortunately, with the onset of the developing world's "debt crisis" two decades ago and the imposition of structural adjustment programs brokered by international financial institutions, these inequalities have widened and deepened.[15]

The Westphalian "map" of southern Africa that emerges from this narrative is one of twelve small, mostly weak juridical entities called "states," which lack the capacity to diversify their economies, develop their human resources, or manage their natural environments. Indeed, only those member states that are offshore—Mauritius and Seychelles—fare any better. However arbitrary and incomplete these data may be, the international community considers this evidence as fact and an immutable context for "development," with the richer states offering "development aid" and financial assistance primarily on a country-by-country basis.

In realist terms, southern Africa's only value is its nuisance value. Influential segments of South Africa's policymaking community are inclined toward this perspective. Hence, South Africa's Ministry of Trade and Industry and Ministry of Finance perceive the need to find a suitable niche within global capitalism, and to continue to dominate the region economically. Similarly, the Ministries of Defense and Foreign Affairs perceive the need to refurbish the armed forces (South Africa's 2000 budget includes a 28 percent increase in defense spending). Denel, the weapons manufacturer, is busy building a global market and forging bonds among these various ministries. Such policy positions only further distance South Africa from its neighbors and reinforce unhelpful binary relationships of "inside" versus "outside" and "us" versus "them."

Within these relatively poor countries, pockets of great wealth are neatly sown into the fabric of grinding poverty. For those SADC countries for which data are available, the gini coefficients (where 0 = perfect equality and 1 = perfect inequality) are as follows: Botswana, 0.54; Lesotho, 0.56; South Africa, 0.62; Zambia, 0.50; and Zimbabwe, 0.57.[16]

Environmental Issues

Mineral- and plantation-based patterns of economic development developed over centuries, and the philosophies that continue to privilege them are at the heart of the region's growing environmental problems. Two issues—freshwater resources and land use—will be highlighted here.

The southern African region is by and large water-poor and prone to drought. Indeed, drought is normal, with regional weather patterns running in roughly eighteen-year wet and dry cycles over the last three centuries. Rainfall in the region is extremely variable, with the north and the east being considerably wetter than the south and the west. Human settlements (and industrial patterns and means of capital accumulation) tend toward the inverse of rainfall patterns—i.e., save for coastal settlements, the higher the rainfall, the lower the population density. This pattern results from human settlements' having grown up around mineral development, most of which is located in arid, semi-arid, and desert areas. A fundamental problem for the region, therefore, is access to freshwater. As populations increase, the question for policymakers seems to be a choice between moving water or moving people.[17]

A related issue involves land use. Approximately 50 percent of the region's land is suitable for cultivation. Much of this land is presently unavailable, however, due to infestations of the tsetse fly. Half of Africa and virtually all of southern Africa is savannah, which is good for maize production. Historically, southern Africa's peoples practiced either slash-and-burn agriculture (*chitemene*), wherein fields would be used for two or three years and then left fallow for twenty or thirty years, or floodplain agriculture (*molapo*), wherein crops would be planted after flood waters had subsided. Both of these practices are under threat. In the case of *chitemene,* the combination of increased population and structural adjustment program–induced private ownership of land has resulted in overcrowding of marginal lands. As a result, the majority of the region's smallholders are forced to work decreasing and marginally fertile tracts. This practice has led to a region-wide trend toward soil degradation (due to erosion, deforestation, overgrazing, and bush encroachment, and compounded by inappropriate land use). A conservative estimate is that 20 percent of the region's arable land is in need of rehabilitation.

In the case of *molapo,* water management and transfer schemes designed to satisfy the increasing demands of industry, mining, big agriculture (irrigation accounts for an estimated 60 percent of water use in the region), and urban households have seriously altered the natural flow of water in all the region's river basins. The creation of Lake Kariba, for example, has decreased annual flooding in the Zambezi River basin by 25 percent. In the absence of flooding, there has been an increased tendency toward stream-bank cultivation, which in turn leads to erosion, siltation of riverbeds, and increased instances of eutrophication, given lower and slower water flows.

Drought compounds these problems. It is not so much the total amount of rainfall annually that determines the sustainability of resource use practices but, rather, variations in rainfall. A five-minute deluge, for example, can be transferred into sustained, multiyear growth by opportunistic plants used to only intermittent rainfall. As such, the region's small and large farms alike are being encouraged to follow opportunistic livestock-management strategies: maximizing stocks following significant events (like La Niña), but maximizing the slaughtering of livestock during the early stages of drought. In the absence of such a strategy, human practices only exacerbate the negative trends already identified, particularly during drought cycles. For example, increasing concentrations of human populations lead to heightened levels of deforestation and clearance of other vegetation. These practices, combined with inappropriate land use—e.g., concentrated stream-bank cultivation, *chitemene* in confined and over-crowded spaces, overstocking on fragile soils[18]—result in soil degradation, particularly desertification. Hardening of the soils—which occurs equally but for different reasons in urban areas and via water-transport schemes—decreases the capacity of the biota to recharge groundwater and often leads to flash flooding. Flash flooding in turn causes further soil erosion.[19]

The Westphalian State, Resource Capture, and Conflict

Without a doubt, natural resources have been, are, and will always be at the center of violent conflict in the region. John Keegan makes a trenchant observation about warfare in general when he rhetorically asks, "Is it true, then, that the zone of organized warfare coincides, inside seasonal variables, with that which cartographers call "the lands of first choice," those easiest to clear of forest and yielding the richest crops when brought under cultivation? Does warfare, in short, appear cartographically as nothing more than a quarrel between farmers?"[20]

His is another way of framing Crosby's observation: the search for and maintenance of "neo-Europes" has been a violent affair. In southern Africa, colonialism meant local peoples' dispossession of the better agricultural lands, while industrial capitalism, with its hunger for minerals, expanded the meaning of "lands of first choice" and extended the lines of formal warfare. State-making and state-maintaining have been and continue to be about material things. In southern Africa, minerals in particular have made and maintain the myth of the Westphalian state.

Unlike the states established at the Treaty of Westphalia, southern Africa's states were imposed by imperial powers. From the outset, "foreign" and "development" policies in southern Africa were made by and for foreigners: Dutch, Portuguese, British, German, and Belgian. As high-lighted above, this type of policy orientation served to disarticulate extant and historical regional relations and disunite the region's peoples in ways previously unimagined. "Independence" and the acceptance of colonial boundaries only served to reinforce the foreignness of policy decisions: in every case, neocolonialism followed colonialism. The fiction of the West-phalian state system in southern Africa contrasts with the reality on the ground: goods, people, resources, and animals continually ignore or seek to circumvent these borders. As will be seen below, these contrasting percep-tions of regional space often lead to conflict.

State-Building?

The practice of Westphalian state maintenance and foreign policy has been highly destructive in the region, not merely in terms of South Africa's regional policy of "destructive engagement,"[21] but more fundamentally in terms of entrenching regional elites who have held jealously to state sovereignty and the perquisites of power attached thereto. Foreign policy, then, has rarely been about people or fostering human security; it has mainly been about material things fostering elite continuity. In Thomas Homer-Dixon's framing, southern African inter- and intrastate relations center on *resource capture* in the context of *structural resource scarcities.*[22]

Historically, it can be seen that minerals have created borders: oil in Angola, copper in Zambia, diamonds in Botswana, diamonds and gold in South Africa. The more valuable and abundant the mineral, the better state-makers have been at maintaining inherited state "forms" in the region. Hence, countries such as Lesotho, Mozambique, Tanzania, and Malawi, with their little-exploited or unavailable mineral wealth, have more difficulty sustaining the claim to sovereign statehood than those previously mentioned. Indeed, even in cases such as the DRC and Angola, where the ability of the government to actually govern within its territorial boundaries is limited and contested, mineral wealth allows state-makers to continue to act as though the Westphalian state were a reality, or to use Robert Jackson's well-known terms, as though they were de facto and not mainly de jure nation-states.[23] In each of these cases, minerals have made borders and entrenched elites but—as human development indicators demonstrate— have had marginal positive impact on the lives of most people.

The cases of Botswana, Namibia, South Africa, and Zimbabwe differ from those of their other SADC counterparts. It is not unreasonable to argue that active programs of "nation building" are underway in each of these four states (recognizing, of course, that they are in fact "state-nations," thus making the projects highly problematic).[24] Capitalism has penetrated more deeply in South Africa and Zimbabwe than it has anywhere else on the continent. As a result, civil societies are active and relatively empowered. State-makers ignore the interests of these civil societies at great peril. Botswana is a country of vast mineral wealth headed by a well-entrenched authoritarian and conservative elite that sits astride a docile and small population; civil society is weak. Namibia, like Botswana, has vast mineral wealth and an entrenched elite. Like South Africa, it is racially divided, has come late to political independence, and so must respond to the demands of a black majority keen on social justice. In each case, the Westphalian state form seems more of a possibility, whether the benefits of "sovereignty" are articulated in the interests of the few or those of the many. State-makers act on this possibility, taking every opportunity to reinforce sovereignty. This is most clearly seen in the Botswana-Namibia case study below.

With the Treaty of Berlin, water has served the process of state-building primarily as a boundary. The Zaire, Cunene, Cuito, Zambezi, Ruvuma, Limpopo, and Orange rivers and Lake Malawi help divide the region into juridical states. Thus, the Cunene and Orange rivers mark Namibia's northern and southern borders (with Angola and South Africa, respectively). Yet, in the space between these two border-marking river systems, there is not one permanently flowing river. Namibia's urban centers evolved out of mineral developments; copper, lead, vanadium, zinc, cadmium, silver, pyrite, and uranium mining are the primary activities in Tsumeb, Grootfontein, and Windhoek. Groundwater, in the meantime, "is not found in abundance where it is required."[25] Similarly, Harare, Bulawayo, and the Johannesburg-Pretoria-Vereeneging industrial triangle are all located far from adequate freshwater resources.

As populations increase across the region, SADC state-makers are being forced to rethink their regard for water as a boundary and for water allocation based on outdated conceptions of riparian rights. Still, the environment broadly defined seems most centrally conceived of as an interlocking series of renewable and non-renewable resources to be exploited for purposes of economic development. So oil and water *do* mix: each is seen to be fundamental to "national security," inclining policymakers toward capture and defense of these resources. More integrated understandings of the environment are continually shunted to the sidelines. Clearly, choices about forms of interstate cooperation regarding the allocation, use, and management of natural resources are being made. As will be seen in the four cases presented below, choices and practices are not always consistent.[26]

Sites and Patterns of Conflict: Between Kant and Clausewitz

Two dominant forms of conflict are extant in the region. The first is a consequence of poverty; the second is motivated by questions of state-making and maintenance. Too often this second form translates into defense of elite privilege and involves concerted attempts at resource capture. Not to downplay the importance of the former, but it is the latter that will be emphasized here.[27]

The southern African region is presently awash in organized forms of violence. These conflicts occur, for the most part, at the substate level, but they regularly involve other states, corporations, and individuals both within and beyond the region. Flashpoints, in descending order of magnitude, include Angola, the DRC, Lesotho, and Botswana and Namibia. Each conflict seems centrally concerned with resource capture as a means to maintain both the current state form and its elite. Each of the four cases will be discussed here, with special emphasis given to the cases of the DRC, Botswana, and Namibia.

Angolan Impasse

Angola appeared close to peace several times during the 1990s, yet each time the warring parties seemed ready to reach an agreement, fighting would resume.[28] Resource capture facilitated this unfolding process. Oil from the Cabinda enclave, territorially separate from, but juridically belonging to Angola, allowed the present Popular Movement for the Liberation of Angola (MPLA) government of President José Eduardo dos Santos to remain in power, to wage war against Jonas Savimbi's National Union for the Total Independence of Angola (UNITA) rebels, and to maintain the fiction that Angola is a viable Westphalian state form. At the same time, the vast diamond fields once controlled by UNITA ensured that organization's capacity to attract outside interest (e.g., in the form of eastern European arms suppliers and mercenaries, or Israeli diamond dealers), to maintain and exercise power, and to contest for control of the Angolan state. Unlike poverty-related, subnational forms of anomic violence, violence in defense of elite privilege, particularly where such highly prized resources are involved, is not so easily contained.

As with the case of the DRC, Angola's neighbors helped prolong the civil war. In line with U.S. Cold War policy, the DRC's former ruler, Mobutu Sese Seko, provided logistical and territorial support for UNITA. In the post–Cold War era, he acted as a conduit for "blood diamonds," the sale of which fueled war and other forms of violence. State-makers in Zambia and the Republic of Congo, also keen to capture some of Angola's vast mineral wealth, provided similar lines of trade and travel for both resources and rebels.

Following a strict policy of military victory without compromise, the dos Santos government succeeded in reducing the once-mighty UNITA to a ragtag group of hit-and-run bandits. UN sanctions and a global movement

against trafficking in blood diamonds helped squeeze UNITA economically. Moreover, Angolan military power was brought to bear in ousting
hostile governments in the DRC and the Republic of Congo, in defending
the embattled DRC government of Joseph Kabila, and in intimidating
Zambia into ceasing its support for UNITA.

With the recent killing of Savimbi, the prospects for peace in Angola
have never looked better. But what is to be the form and content of this
"peace"? Clearly, the absence of conflict over much of territorial Angola
would be welcome. However, the ability of the MPLA government to
"broadcast" its power will remain limited to the capital, Luanda, and its
environs.[29] Much of Angola will remain beyond Westphalia, and resource
capture will continue to determine both the ability of the MPLA to rule and
the forms of resistance likely to emerge in the countryside.

Water as Power: The Case of Lesotho

Lesotho, unlike Angola, is often held up as an (ongoing) example of successful regional cooperation in the areas of resource use, management, and
conflict resolution. In the case of resource use, the Lesotho Highlands
Water Project (which involves a massive diversion of the tributaries of the
Senqunyane/Orange River) is cited by state-makers in South Africa and
Lesotho as a prime example of the possible benefits to be had from interstate cooperation on the sharing of scarce water resources. In terms of
conflict resolution, Lesotho has been the "test case" for SADC's Organ on
Defense, Politics, and Security. In 1994, Lesotho was the object of an
informal intervention by SADC's "three M's"—President Mugabe of Zim-
babwe, President Nelson Mandela of South Africa, and President Ketumile
Masire of Botswana (who was replaced as president by Festus Mogae in
1998)—who were called upon to intervene in Lesotho on behalf of the
duly-elected Basotholand Congress Party (BCP) government of Ntsu
Mokhele. Mokhele had been deposed in a constitutional coup staged by
King Letsie III with the aid of the highly politicized, anti-BCP Lesotho
Defense Forces (LDF).[30]

Having successfully convinced the king to reinstate the BCP government, SADC leaders congratulated themselves on the new regional practice
of "defending democracy." More recently, SADC was called to intervene in
Lesotho once again on behalf of a duly-elected government. This time,
however, force was used, combining troops from South Africa and Botswana. More than fifty Basotho and eight South Africans were killed in the
fighting.

Of related interest was the firefight that took place between SADC forces and members of the mutinous LDF at the Katse Dam. In Lesotho—a small, landlocked country whose major export was, until recently, labor sent to the mines and farms of South Africa—control of the Katse Dam was tantamount to control of the country's means of production. With the massive retrenchment of Basotho mineworkers, the sale of water to South Africa now constitutes the country's major source of foreign exchange and government revenue. Water, while providing energetic power for South Africa's industrial heartland, also facilitates the political power of state-makers in Lesotho.

The language of cooperation notwithstanding, policymakers in the SADC states are clearly operating from a realist framework. Amid all the saber-rattling, the regional "peace dividend" rings hollow. More seriously, the framing of peace through interstate diplomacy helps reinforce the image of Southern Africa as a collection of self-regarding and independent states. In cases such as Lesotho, it allows weak states and selfish state-makers to wallow in dysfunctionality. In the case of Namibia and Botswana, to be discussed below, it helps reinforce the image of active nation-building.

Resources for Capture: Conflict in the DRC

Clearly, policymakers in both South Africa and Lesotho define water as a factor of production and thus a resource befitting capture. The Highlands Water Project marks the high point of joint exploitation of this resource. In the case of the DRC, interested parties conceptualize the DRC itself as a resource pie ready for taking. State-makers and businesspeople in South Africa, Namibia, and Zimbabwe are particularly interested in gaining long-term access to the DRC's resources. Namibia's state-makers are after water; Zimbabwe's are after mineral resources; South Africans are interested in energy, water, land, and minerals. All three, along with other SADC member states, are interested in a politically stable and economically prosperous DRC.

It is estimated that the Inga rapids, located 30 kilometers (km) upstream of the Angolan border, represent the largest single hydroelectric energy potential in the world. According to Alan Conley of South Africa's Department of Water Affairs, Inga has a potential generating capacity of 45,000 megawatts (MW). This may be contrasted with the 4,500 MW presently generated in the Zambezi basin at Kariba, Cahora Bassa, Kafue Gorge, and Nkula. Given South Africa's desire to develop "clean" and renewable means of electricity generation, the prospect of developing and linking Inga into a region-wide energy grid is most appealing.

Whereas much of the SADC region suffers from water stress, the DRC is a water-rich region, with mean annual precipitation (MAP) of 1,500 millimeters (mm). This amount may be contrasted with South Africa's MAP of less than 500 mm. Although South Africa accounts for a mere fraction of the renewable water resources in the region, it consumes more than 80 percent of them. Moreover, it is estimated that by 2020, South Africa, Botswana, Malawi, Namibia, and Zimbabwe will have entered conditions of absolute water scarcity. If predictions of global warming prove correct, the twenty-first century will be hot and dry for much of southern Africa. The region, therefore, will continue to look for ways to satisfy its growing thirst. It is interesting to note that rather than seriously considering demand management of existing water supplies, state-makers are instead keen on accessing water from the Congo River.

The Congo River basin covers an area of 3.98 million km^2 and has a mean annual runoff of 1.25 billion cubic meters (m^3)—twenty-five times the mean annual runoff of all of South Africa's rivers combined, more than ten times that of the Zambezi River basin, and more than one hundred times that of the Okavango River basin. Although the logistics of transferring water southward from the DRC are daunting and speculative at best, it is clear that the DRC's water resources constitute a primary reason for drawing it into the SADC web of protocols, in particular the protocol on shared watercourses and energy development. The Congo basin encompasses Burundi, Rwanda, the Central African Republic, Cameroon, the Republic of Congo, and SADC member states Tanzania, the DRC, Zambia, and Angola. At the same time, Namibia's leader, Sam Nujoma, has given the go-ahead for discussions regarding the possible construction of a pipeline stretching from the mouth of the Congo River to Namibia's capital, Windhoek.

Land is also a critical factor. Although only 3 percent of the DRC is classified as arable land, 74 percent is forest. The environmental lobby in South Africa is gaining strength and there is a great deal of resistance to the further expansion of commercial forestry. Pulp and paper giants such as Sappi are accused of environmentally destructive practices. For example, introducing water-sapping exotic tree species (primarily pine and eucalyptus) to upper catchment areas of principal rivers has a severe impact on downstream users, as flows are reduced and sediment loads increased. Moreover, as industry concentrates on the production of pulp and paper for export, these forests contribute little or nothing toward satisfying domestic demand for fuel wood and building materials. Should forestry law change in South Africa, the pulp and paper giants view the DRC as a potential new area for exploitation.

At the same time, the DRC represents a potential means for defusing conflict over land within South Africa. Numerous coffee, sugar cane, palm oil, and rubber tree plantations dot the south of the DRC, suggesting scope for expansion. At present, many white South African farmers are moving into Mozambique and Zambia, having sold their farms to the government in anticipation of a prospective land grab. This "second trek" is being facilitated by bilateral agreements between governments in the region. Similar agreements may be reached between South Africa and the DRC, facilitating a trek further north.

As with land and water, so with factors of production under the land. In the southern part of the DRC are large deposits of copper, cobalt, tin, zinc, and uranium, among other minerals. South Africa–based multinational corporations such as Anglo American and De Beers are keen to exploit these resources. However, then-president Laurent Kabila accused both of these corporations of "monopolism" and of "raping" the DRC through the extraction of mineral wealth. There are reports that, as far back as July 1997, Harry Oppenheimer, the recently deceased patriarch of Anglo American, telephoned Mandela asking him to intervene on behalf of the giant mining house. At the same time, Nick Segal of the Johannesburg Consolidated Investment Co.—South Africa's first black-controlled mining house—put in a call to Deputy President Thabo Mbeki. This resulted in an initial meeting between the government, the mining houses, the giant electricity parastatal Eskom,[31] and other elements of business in South Africa to discuss a way forward in the DRC.

Building Regional Peace with the Factors of Production?

Along with resources, regional stability is also a critical consideration. If current South African president Thabo Mbeki's hypothesized "African renaissance" is to have the slightest chance of being realized, it hinges on a stable central Africa.[32] In this context, South African and other policymakers in the SADC states continue to pursue a twin agenda in the DRC: containment and development. In the case of containment, the SADC members fear that a collapse of the state in the DRC will lead to a massive flood of refugees across their borders. This prospect further fuels not only fears of a drain on state resources but also xenophobic worries of Acquired Immunodeficiency Syndrome (AIDS) and Ebola pandemics. In the case of development, then, by bringing the DRC into SADC, policymakers hope to preempt collapse by active involvement in the reconstruction and development of the DRC. Indeed, Laurent Kabila indicated his state's desire for such help. As a regional project this manifests itself most clearly in efforts to link the DRC to a program of shared energy development via the Southern African Power Pool (SAPP).

The SAPP is driven by the power utilities of the region, with support from the SADC Energy Sector's Technical and Administrative Unit (TAU). The power pool was formalized first through the signing of an intergovernmental memorandum of understanding in December 1995, and second in the SADC energy protocol signed a year later. The DRC's Société Nationale d'Electricité has been involved with the SAPP from the beginning. The SAPP's main aim is to foster and develop a regional electricity infrastructure of mutual benefit to member states. According to an Eskom report, the SAPP hopes to create mechanisms by which member countries can trade electricity among themselves, in particular between those in the north who have vast hydroelectric resources and those in the south who have large loads and thermal plants. To achieve this, high-capacity electrical transmission lines must be built, interconnecting the various national networks. In the longer term, the success of the grid will be realized when major hydro potential on various rivers is used to send substantial amounts of electricity to other African countries. The Congo River figures centrally in these plans.

Many potential benefits are to be had from regional energy cooperation. First, creation of a region-wide energy grid would allow for better economies of scale, minimizing costs and maximizing availability. Second, conjunctive use of hydropower facilities, in combination with an optimum mix of power supply sources, would enhance efficiency, reliability, and ultimately regional energy security. Third, by drawing South African technical expertise and (multinational) corporate financing into the energy development picture, the region may begin to exploit new and renewable sources of energy, thereby avoiding further adverse environmental impacts. Fourth, by advancing this sort of functional regional cooperation, habits of acting transnationally may be developed, which in turn provide a platform for further regional integration.[33]

Among SADC's ten original members—i.e., the land-based states excluding South Africa—electricity generation is primarily realized through hydropower. Installed capacity is an estimated 10,000 MW, with potential for another 8,000 MW in the Zambezi basin alone. Peak loads represent less than one-half of installed capacity, so there is significant scope both for expanding delivery into rural and peri-urban areas and for exporting to other African countries. Eighty percent of this installed capacity emanates from Zimbabwe and Zambia. The DRC is connected to the grid via Zambia, a link first established in 1956. Tanzania, Malawi, and Angola are not as yet connected to the grid. Most other electricity parastatals operate at significant losses due to diseconomies of scale.

The SADC states hope that the establishment of a regional energy grid will decrease costs and increase the reliability of supply. Together, these two benefits will enhance regional options for economic development. Power continues to be seen as the necessary means to industrialization and diversification of economies overly dependent on primary products for export and saddled by the high cost of fuel imports.

At the same time, it is hoped that cheaper, reliable sources of hydropower will assist the "fuel transition" in SADC countries. Unlike most of the rest of the world, urban areas in SADC are experiencing "backward fuel switching" instead of a continuing transition to modern fuels. That is to say, as electricity supplies continue to be unreliable, urban dwellers increase their consumption of traditional sources of energy. The energy crisis in SADC's original member countries, then, is a crisis of dwindling supplies of fuel wood. An estimated 95 percent of SADC households are dependent on biomass—in particular coal, charcoal, and firewood—for their energy needs. To avoid increasing deforestation, and to lessen the burden on women and children in the gathering of fuel wood, electrification projects in rural and peri-urban areas are a central aim of the SAPP.

Energy in South Africa: The Search for Cleaner Sources

In South Africa, the attraction of hydropower in general and the DRC in particular involves two issues: potential contributions to the Reconstruction and Development Program (RDP) and trade considerations. South Africa's energy profile reflects the decades of isolation and sanctions the country suffered as a result of apartheid. Energy for survival, let alone development, resulted in an industry motivated by the drive for self-sufficiency, shrouded in secrecy, and for which economic and environmental costs were not priority issues. South Africa accounts for an estimated 90 percent of the total coal reserves in Africa. Coal-based energy development therefore became the key to apartheid South Africa's survival and industrial develop-

ment. The result today is a sector that is not sustainable: it is dirty, economically inefficient, energy intensive, and caters to a narrow band of the
population.

South Africa's estimated energy consumption per unit of GDP is 3 times
that of the United States, 4 times that of the United Kingdom, 5 times that of
Germany, 5.5 times that of France, and 7 times that of Japan. South Africa
is the twelfth-largest producer of carbon dioxide in the world, behind
France—even though France has a GDP 11 times greater than South Africa's. Whereas South Africa accounts for 0.7 percent of global population,
it accounts for 2 percent of carbon dioxide emissions.

Eskom has an installed capacity of 38,000 MW, centered around a series
of 3,600 MW stations in the high veld of the province of Mpumalanga
(formerly Eastern Transvaal). Eskom is one of the top seven electricity
producers in the world, accounting for more than 50 percent of all electricity generated in Africa. There is significant overcapacity in South Africa
at present, and although domestic consumption of electricity accounts for
only 15 percent of total energy use, it is prone to peaks, especially during
the hot summer months and extremely cold winters.

Because of the necessities of survival in the face of international sanctions, energy supplies in South Africa grew to be highly subsidized, cheap,
dirty, and inefficient. Given the low cost of energy to industry, there was
little incentive to clean up, conserve, or make it cost effective. There are
pressures for this to change, however.

As in the rest of the SADC region, access to electricity is extremely
unequal. An estimated twenty-five million people in South Africa have no
access to electricity and 50 percent of the population relies on fuel wood for
its energy needs. Low-income homes, particularly those located in the
townships, rely heavily on coal, kerosene, paraffin, charcoal, and wood for
their energy needs. As the highest-grade coals are exported, both industry
and individuals in South Africa consume low-grade coal that is low in
sulphur content but high in ash. This trend contributes significantly to health
problems among poor households and to respiratory diseases, in particular.

The terms of the RDP indicate an intention to provide affordable electricity to rural and township areas in South Africa. Granted, this electricity
will not immediately displace coal, as coal remains cost-competitive and
the cost of electrical appliances is very high. Nevertheless, schemes such as
that undertaken in Khayelitsha, the large township outside of Cape Town—
where Eskom, in cooperation with EDF–Southern Africa, a division of the
giant French parastatal Electricité de France, provided low-cost electricity

to an estimated 43,000 homes and 500,000 people—suggest that delivery is both possible and welcome. Access to electricity is based on a prepayment, code-based metering system. Eskom is involved in similar schemes in Tanzania. In the short term, Eskom argues that the transfer of pollution out of the townships and into the Mpumalanga high veld is a reasonable option. If combined with the retrofitting of de-sulphurization technology, increased demand for coal-fired electricity will not contribute markedly to environmental pollution.

In a government-commissioned study, the Development Bank of Southern Africa estimated that a total of 5.6 billion rand worth of "energy intensive" goods was exported to "environmentally sensitive countries" in 1994. Given the legacy of cheap, dirty, and energy-intensive production processes, there are worries that South African exports will be penalized under emerging transnational "eco-labeling" legislation. If eco-labels are to include either "energy per unit of production" or "externalities associated with production" in their assessment criteria, South African goods will indeed be subject to penalty. Production of gold and aluminum, for example, produces a great deal of toxic waste. And, since all South African goods are dependent to some degree on the burning of low-grade coal, each product could be subject to international sanction. As with the RDP, therefore, trade considerations highlight the need for, and push toward, cleaner and sustainable sources of power.

Beyond Factors of Production: In Search of Ecosystems and People

All told, there are myriad pressures pushing South African and SADC state-makers toward increased involvement in the DRC in general and the Congo River basin in particular. Yet there are extant worries about the social costs of energy development. The unsuccessful relocation of the Tonga peoples from the Gwembe valley during the construction of the Kariba Dam is a persistent wart on the face of regional development. And the Tonga are again in the news: with Zimbabwe's decision to push ahead with the Batoka Dam hydroelectric scheme, the Tonga will once again have to be moved. Similar difficulties have been experienced with the displacement of villagers from their traditional homes in the Lesotho highlands to make way for the Katse Dam.

The social and environmental costs of dam development are well known, yet these giant hydroelectric and water-transfer schemes continue to be favored by state-makers keen on ready sources of electricity and hence revenue generation. On the positive side, however, there are few today who would regard Lake Kariba as a liability rather than an asset to the region. The rehabilitation of the Cahora Bassa dam in Mozambique is regarded with similar excitement and anticipation. Moreover, the simple fact is that big business and big government are going to continue to drive "developmental" processes in the region, no matter how destructive of the environment or devastating to the disempowered. The key is to push for a transparent and broadly based decision-making process, an issue that will be discussed further below.

Eskom's demonstrated capacity for delivery, based on its experience, capital, and human resource advantages, means that it is the driving force behind regional energy production and management. Given the legacy of ineptitude among other parastatals in SADC states, this is a positive development. Moreover, there are important, positive pressures pushing Eskom into the region: the search for cleaner and renewable sources of energy, the prospect of cheaper and more reliable sources of electricity for all SADC states, the provision of electricity to previously disadvantaged sectors of society, and the necessary infrastructure development in a long-devastated non-economy like the DRC, to name several.

If balanced economic growth, equity of access, and environmental sustainability are to be at the center of energy development, it is imperative that the energy sector in South Africa abandon its penchant for secrecy. Transparency and grassroots involvement are fundamental to this process; they will also help ensure that the narrow interests of multinational capital and state leaders do not always win the day. The World Commission on Dams (WCD), chaired by South Africa's former minister of water affairs and forestry, Kader Asmal, is a step in this direction. Although dominated by technical perspectives and expert knowledge regarding water transfer and impoundment, the WCD has provided something of a forum for other, marginalized voices to be heard. At its February 2000 meeting in Harare to present findings on the Kariba Dam case study, twenty chiefs from the displaced Tonga made formal statements. They were assisted by international nongovernmental organizations in gaining this platform, a distinct improvement from the one member of the community invited to the WCD council meeting in Cape Town in April 1999. At the same time, it is important that hydropower and new sources of supply not be regarded as the panacea for the region's myriad ills.

As with the situations in Northern Ireland and the Middle East, bringing peace to the DRC is a complicated and never-ending process. The intransigence of all concerned reflects the size and value of the natural resources to be captured.

Botswana and Namibia: The Case of the Okavango Delta

In Botswana and Namibia, the issues are smaller, they involve the use and management of a common resource, and, though centered on overlapping ecological interdependencies, conceptions of appropriate use are based on fundamentally opposed knowledge claims.[34] The issues referred to are, first, Namibia's desire to draw water from the Okavango River system, and second, competing sovereignty claims regarding three small islands in the Chobe River.

At first glance, the fears and conflicts regarding water use in the Okavango River basin seem misplaced. Botswana and Namibia are signatory states to the 1997 UN Convention on the Law of the Non-Navigational Uses of International Watercourses and the Helsinki Conventions; they established a Joint Permanent Water Commission in 1990; and they, along with Angola, established OKACOM in 1994. In addition, the SADC protocol on shared watercourse systems, which initially developed out of discussions regarding activity in the Zambezi River basin, was signed by most SADC countries at that organization's heads of state summit in 1995. It has since been ratified by a two-thirds majority of SADC states and is now binding. Fundamental to proposed developments in the Okavango River basin is the following principle from that protocol: "Member states lying within the basin of a shared watercourse system shall maintain a proper balance between resource development for a higher standard of living for their peoples and conservation, and enhancement of the environment to promote sustainable development."[35]

In addition, the Okavango delta was declared a World Heritage Site in 1997; as such, it is in the global interest that its ecology be preserved.[36] At present, there are very few violent human interventions regarding water abstraction in the basin. In the catchment area of Angola, there has been no development since the outbreak of war more than twenty-five years ago. (Angola contributes more than 90 percent of the mean annual runoff of the river system.) Namibia pumps water for irrigation in the northeast, having a small and very inefficient dam at Omatako. In Botswana, the sole intervention is the Mopipi Dam at the diamond mine-head in Orapa. This relative lack of development may be compared with the thirty-two large dams in the four countries in the Zambezi River basin.

Namibia has considered pumping water from the Okavango to its water-poor capital, Windhoek, since 1969. Indeed, the Namibian National Eastern Water Carrier (NNEWC), started at Windhoek and existing in several different phases of completion, has been under construction for thirty years; it presently extends as far as Grootfontein, some 250 km from the intended abstraction point at Rundu on the Namibian border with Angola. The proposed off-take is a modest 1 percent of the river's total annual flow, or approximately 100 million m³ per year.[37] Off-take will be via an abstraction tower so as to minimize disturbances to the movement of sediment in the river.[38] It is estimated that a 1 percent off-take will reduce the delta's floodplain by 3 percent, possibly affecting the local population's dependence on *molapo* agriculture.

Botswana also considered diverting water from the delta at the Boteti River, which flows as an offshoot from the swamps south of the main settlement at Maun. This plan, known as the Southern Okavango Integrated Water Development Plan, was shelved in 1992 partly in response to severe criticism by the World Conservation Union (IUCN), which was contracted to conduct the environmental impact assessment. The IUCN suggested that different ways of providing water to the rest of the country—e.g., via demand management—be explored first. The primary impetus for shelving the project, however, came from a hastily organized civil society: more than three hundred chiefs in the Okavango delta region signed a petition opposing construction of the dam. It is this legacy of an organized social movement arising in response to a perceived crisis that provides hope for longer-lasting forms of subnational and possibly transnational development in the Okavango region. This point is discussed further below.

In contrast to the Boteti project, the impact assessment undertaken for the afore-mentioned NNEWC project found it to be sound. This, then, marked the beginning of a conflict-laden interstate discourse over resources in the delta region. Although the manner in which the issue is playing itself out mirrors conflicts over land and water throughout the SADC region, the case does have hopeful aspects.

Water as Power

The government of Botswana has found Namibia to be most helpful in the former's state-building project. To manufacture unity from within, state-makers have often manufactured a threat from without, and Namibia's intention to draw water from the Okavango is articulated in Botswana as a threat to "national security." Their focus has been on events in the Okavango and also the Chobe-Linyanti river systems. In the case of the former, much is being made of Namibia's plans to go ahead with construction of the NNEWC pipeline. The media and international NGOs have contributed to this discourse by issuing an endless stream of alarmist journalism. Recent headlines appearing in local newspapers have included "Plan could turn Okavango to dust," "Namibia almost certain to drain Okavango," "Will the Okavango nightmare come true?" and "War Clouds."

Interestingly, state-makers in Botswana who have long been at odds with international NGOs over environmental issues presently find themselves literally on the same side of the river. They are actively working with several local and international NGOs, such as Conservation International, the International Rivers Network (IRN), and the Kalahari Conservation Society—who together with the University of Botswana make up the Okavango Liaison Group (OLG)—to devise a strategy for confronting the Namibian government on its proposed off-take from the Okavango River. Recently, the OLG was empowered by a coalition of fifty chiefs and community leaders in the delta to draft a petition on their behalf demanding that the NNEWC pipeline project be stopped.

As with the Boteti Dam case cited above, this petition was signed by more than seven hundred influential people in the delta region. The petition itself was initiated by the IRN, but ownership of the issue is not seen to accrue exclusively to an international NGO. To the contrary, the IRN has worked very hard, through the office of the OLG, to ensure that coordinated action on the pipeline issue rests on a broad-based consensus of regional stakeholders. Indeed, with seed money from the Swedish International Development Agency and coordination from Botswana's Kalahari Conservation Society, the OLG has recently begun to explore possibilities for transforming itself from a single-issue social movement into a transnational, regionally oriented, sustainable development program. Granted, the delta is a very large area with diverse interests at stake. However, the OLG, now renamed Every River Has Its People, hopes that the energy generated in opposition to the proposed NNEWC pipeline can provide an impetus for more broadly based regional cooperation.

Post-Westphalian Governance?

For the purposes of this chapter, it is prudent to attempt to deconstruct the discourse around the proposed NNEWC. As it stands, the conflict between Namibia and Botswana over the NNEWC rests on rival knowledge claims: in Namibia's case, that the proposed off-take will not affect the ecology of the swamps; in Botswana's case, that any off-take will set the precedent for future, perhaps more significant, off-takes, particularly in Angola, and so will lead the delta down the slippery slope of destruction. Each side has marshaled its own experts to verify the sustainability of desired national policy outcomes.

In addition, given that both Namibia and Botswana are liberal democracies, policymakers are seeking to speak to and placate different audiences with widely varying interests. Namibia, like Botswana, suffers from an historical "race-space" division of land. The 60 percent of the country's population that lives in the northern communal areas (see Map 5.1) occupies land of low quality. Only 20 percent of Kaokoland, 24 percent of Ovamboland, 7 percent of Okavangoland, and 52 percent of the Caprivi Strip is suitable for agriculture (and this is both a small, narrow band of land and one prone to infestation by the tsetse fly). Although whites make up no more than 8 percent of the country's population, they control 60 percent of

Map 5.1
Land Distribution in Namibia

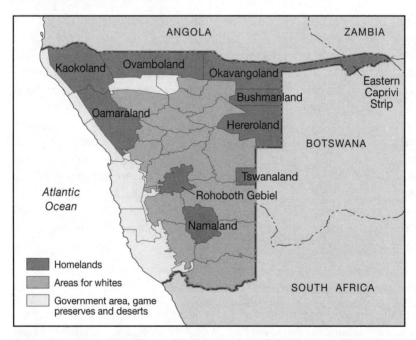

Source: S. Iremonger, C. Ravilious, and T. Quinton, eds., *A Global Overview of Forest Conservation, Including GIS Files of Forests and Protected Areas,* ver. 2 (CD-ROM) (Cambridge, U.K.: Center for International Forestry Research and World Conservation Monitoring Centre, 1997).

the agricultural land, almost all of which is devoted to pastoralism (sheep and karakul production). This land accounts for 80 percent of the total agricultural yield. To be sure, since independence in 1990 there has been a marginal influx of black farmers into formerly white-held land. Like those in South Africa and Zimbabwe, however, these "farmers" are mainly drawn from the new ruling elite. The average freehold farm size is 7,200 hectares, whereas the average holding in communal areas is 17 hectares. Moreover, because these lands are communally held, there is no guarantee of tenure or exclusive use of farm or ranch plots.

Though there is acute scarcity of land in the communal areas, a recent commission resolved that "present communal areas should be retained, developed and commercialized where possible and expanded where necessary."[39] Some effort has been made by state-makers to develop agriculture, particularly in the Eastern Caprivi Strip via the provision of irrigation, state subsidies, and extensive programs of spraying the pesticide DDT (dichloro-diphenyltrichloroethane) to reduce instances of tsetse fly invasion. (The Eastern Caprivi Strip borders on the Chobe River and marks the site of conflict with Botswana over island ownership; it will therefore be dealt with in more detail below.)

Thus land-use patterns in Namibia reflect both race and class prejudices. Commercial farms are held by white settlers and a small but expanding black political elite. In many instances, these holdings are underutilized. In the communal areas, natural population growth and the return of refugees from the war of independence led by the South-West African People's Organization have contributed to acute soil degradation, deforestation, and an imminent fuel-wood crisis. Land clearing along the Okavango River for millet production has, according to one report, "left nothing but exposed surfaces."[40] In addition, the migration of large mammals—wildebeest, elephant, buffalo, and zebra—exposes peasant farmers to livestock death (via the contraction of diseases such as cattle lung) and crop damage. Namibian law permits the hunting of a wide range of wildlife—eland, kudu, gemsbok, duiker, bushpig, blue wildebeest, springbok, and buffalo— but excludes those that tend to do the most damage. Life in the rural areas that border on Botswana, therefore, is very harsh, despite numerous promises by Nujoma's government to help ease the burden via, among other things, land redistribution and the provision of water resources.

In sum, Namibia's state-makers pursue policies aimed at a dual constituency: those involved in the formal economy and those involved in peasant, largely subsistence agriculture. The former cluster around the mines, are big commercial farmers and fishers, and are for the most part of German or

South African origin. Though they are small as a percentage of the population, they are politically and economically powerful. The latter live in the homelands, are peasant farmers or hunter-gatherers, and are indigenous to the region. They are large in number but wield virtually no economic power. Politically, however, these peasant farmers or hunter-gatherers are a potent force: in Namibia's nascent democracy, their consent—either through the ballot box or in passive nonparticipation—counts. Recently, an Eastern Caprivi separatist group has emerged, which suggests that all is not well among this constituency. If elite interests in maintaining political power are to be served, neither the rural poor nor agro-industrial capital can be ignored. Both groups need water—for farming and pastoralism, industry and mining, and household use—which suggests possibilities for compromise.

As can be seen in Table 5.2, these two constituencies have very different water use profiles. If one uses the ratio of water use to gross national product (GNP) contribution as a measure of the relative values of the uses of scarce water resources, it becomes clear that cattle and irrigation represent very low-value uses. Yet, if Namibia's elite is to make good on its election promise to help the majority of rural poor lead better lives, the government will clearly be pressed to provide more water for cows and farming, despite these low use-values. In short, irrigated agriculture and cattle keeping may not pay their way, but they reflect and support a way of life, and the government must therefore act in their support. At the same time, it is difficult to make a case for increased investment in tourism or mining,

Table 5.2

Water Use and Contribution to GNP in Namibia, by Economic Sector, 1997

Sector	Water use (million cubic meters)	Percentage share of water use	Percentage share of GNP	Ratio of GNP share to water share
Irrigation	107	43	3	0.07
Cattle	63	25	8	0.32
Household	63	25	27	1.08
Mining	8	3	16	5.33
Tourism	1	0.3	4	13.33
Industry and commerce	7	3	42	14.00
Total	249	100	100	

Source: John Pallett, ed., *Sharing Water in Southern Africa* (Windhoek: Desert Research Foundation of Namibia, 1997), p. 102.
Note: GNP = gross national product.

as these two activities are capital intensive, employ very little local labor, and are dominated by foreign interests. Moreover, in the case of tourism, as will be seen in the case of Botswana below, the emphasis on "low volume/high cost" tourism, which is good for the preservation of delicate ecosystems, tends to breed animosity among local peasants who continue to view "conservation" as exclusivist and racist.

Furthermore, given Namibia's commitment to "modernizationist" forms of development—in fisheries, mineral beneficiation, and intensive cash-crop agriculture, for instance—the country's demand for water will increase dramatically in the twenty-first century. For all these reasons and more, it is clear that the proposed NNEWC must go ahead. It should be noted, however, that policymakers in Namibia foresee this new supply of water as one element in an integrated management system combining elements of demand management, recycled water use, right pricing, and so on.

In Botswana, the Okavango delta is home to an equally bewildering array of stakeholders, all of whom are interested in the preservation of the wetland in its present form. Local and international NGOs are interested in both conservation and sustainable utilization. They often tie their interests to indigenous communities, such as the San. Along with a wide variety of international donors, these local and international NGOs are at the forefront of developing community-based programs for natural resource management in the delta. These activities vary from joint management of safari concessions to the commercial development of veld products. Farmers are also very active and numerous in the delta.

The delta is home to three forms of land tenure. It is mostly ringed by tribal territories and areas that are freely open for grazing and arable agricultural development. Like in Namibia's homelands, agriculture and grazing land here is communally exploited. The southern edge of the delta is also home to Tribal Grazing Land Policy ranches—i.e., formerly communal territory that has since been privatized. In the northeast, state land predominates with the establishment of national parks and game and forest reserves (see Map 5.2). Cattle keeping is a way of life for all Botswana and marks a cross-class linkage that binds policymaking elites closely to even the poorest of Botswana—who may not now own cows but perhaps will one day. The delta is therefore ringed by a seemingly endless series of fences, designed to separate wild animals from the domestic ones. These fences are highly controversial, as they have truncated the migratory routes of many animal species and have caused death for the millions of animals caught and starved in their barbs. The delta is a regular stopping point along

Map 5.2
National Parks and Game Reserves in Botswana

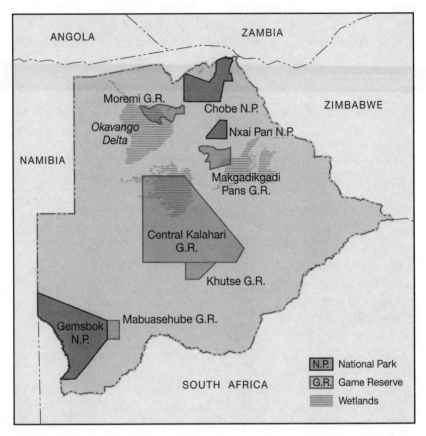

Source: S. Iremonger, C. Ravilious, and T. Quinton, eds., *A Global Overview of Forest Conservation, Including GIS Files of Forests and Protected Areas,* ver. 2 (CD-ROM) (Cambridge, U.K.:
Center for International Forestry Research and World Conservation Monitoring Centre, 1997).

the transmigration routes of elephant, wildebeest, zebra, and buffalo,
among other species.

Tourists and tour operators are keen to see these fences taken down. At
present there are fourteen areas demarcated for various tourist activities:
five territories that are open for general bookings and that include both
trophy and photographic safaris; two national parks (Chief's Island in the
heart of the swamps, and Moremi, which extends north and east to connect
with Chobe National Park); three territories rented annually to safari com-

panies; and four territories that are outright safari company concessions. Although tourists, tourism operators, conservation organizations, indigenous peoples, cattle keepers, and *molapo* farmers are all interested in seeing the delta remain in its present form, they are in no way united regarding present forms of land allocation, use, and management. They tend to split along Aldo Leopold's long-ago hypothesized A-B cleavage, wherein those in the "A" camp conceive of land as a commodity to be exploited to its maximum (cattle keepers who have the traditional ear of the government), and those in the "B" camp regard land as a biotic system (conservation and "eco-tourism" operators, and the San). Others tend to fall somewhere in the middle (safari operators and *molapo* and other peasant farmers). Given the vast capacity for tourism to generate foreign exchange, those in the "B" camp increasingly find themselves sleeping with strange bedfellows, including powerful government members who have purchased stakes in safari concessions around the country.

It is clear that Namibian and Botswana claims of "expert" knowledge rest on fundamentally different conceptions of a river basin: to the former, a basin is simply a channel; in the latter, it is a hydrological unit, with the delta conceived to be the central part of a biotic pyramid. These differing conceptions arise from the popular pressures elites face in making choices regarding land and water use. It is interesting, however, that Botswana policymakers occupy the high moral ground on this issue: preservation of the delta seems to be forcing them to reconsider natural resource use patterns, and to favor sustainable and more equitable uses. Clearly, to statemakers in both Botswana and Namibia, water is power. In the case of the former, however, this narrow, state-centric conception happens to coincide with more sophisticated understandings of water as part of an ecosystem.

Democracy, however shallow, facilitates dialogue both within and between Botswana and Namibia. And relatively new institutional structures—like OKACOM and the SADC water protocol—help guide decisions about natural resource use toward cooperative rather than conflictual outcomes. The fragility of these institutions, however, combined with the problems of Westphalian governance in weak southern African states, cannot ensure that natural resource management issues will not be used for narrow, elite-maintaining political purposes. At the same time, intriguing forms of governance are nascent. For some time now, Conservation International has been working with the governments of Botswana and Namibia to create a trans-Caprivi superpark. This suggests that certain groups within the Namibian government recognize the economic benefits of preserving the delta. This notable Namibia-Botswana cooperation leads us directly to a discussion of the Sedudu island dispute.

Sedudu as State-Building

Until recently, Botswana and Namibia were at the center of a dispute over ownership of a landmass in the Chobe River, a tributary of the Zambezi River. The dispute centered on the island of Sedudu, considered a single piece of land by the Botswana but as two separate islands, called "Lunyondo" and "Kasikili," by the Namibians. A third island, Sitingu, is nearby and is seasonally farmed by Namibian smallholders. Each of these three islands is really part of a greater wetland area in the Chobe River. Indeed, the exposed landmass varies seasonally depending on the rise and fall of the river. The problem emerged when it was discovered that, because of natural processes (e.g., the accumulation of silt deposits over time, and the large hippopotamus population in the wetland area), the deepest part of the channel had shifted. Given the common practice of marking boundaries based on the deepest part of the channel, Namibian state-makers claimed the wetland area as their own. Ultimately, the case was referred to the International Court of Justice (ICJ) at The Hague.

Prior to this referral, members of the Botswana Defense Force (BDF) occupied Situngu, raising the ire of the Namibians. Namibia's minister of defense, Erikki Nghimtina, believed that Botswana should have first consulted with the Organization for African Unity, the UN, or even SADC about how to proceed in the dispute rather than simply occupying the island outright. The BDF claimed no need to clear occupation of Botswana's sovereign territory with anyone. Their view was succinctly stated by Botswana's high commissioner to Namibia, Edwin Matenge: "It is BDF's normal assignment to patrol the borders including the Botswana side of the Linyanti and Chobe [rivers] to curb any illegal cross-border activities, including game poaching and encroaching into Botswana territory through such human activities as plowing, livestock rearing and grazing." Ironically, Namibian villagers from the nearby village of Singobeka farm this supposedly "Botswana sovereign territory" and have done so for years in full view of the BDF.

The BDF's occupation of the island was consistent with plans to erect and occupy camps every 10 km from Situngu to the border located near the Mahangu Game Reserve. Despite reports by an alarmist press, there is very little likelihood that a shooting war would ever develop between these two countries. Indeed, Botswana and Namibia have met yearly since Namibia's independence from South Africa through the auspices of the Namibia-Botswana Joint Commission on Defense and Security. Judging from the communiqués issued from these meetings, the two sides are more interested in jointly combating drug trafficking, poaching, vehicle theft, illegal migration, and other forms of cross-border crime than they are in confronting each other. This increased military presence in and around Chobe National Park in the northwest of the country dovetails with Botswana's determined military buildup (an activity that has raised much concern in the SADC region).

Namibia's foreign minister, Theo-Ben Gurirab, alleged, "We've gone through war and destruction. The last thing we should be spending money on is military equipment." Botswana's then-commander of the BDF, Major General Ian Khama (who is now vice president and minister of presidential affairs) was more forthright. Botswana, he said, acquired "no army, no infrastructure, nothing" at independence. "Now, eighteen years on, . . . we are developing . . . a force that will not see our security compromised." Like their South African counterparts, policymakers in Botswana openly acknowledge their fears of regional "insecurity." While on an official visit to South Africa, Khama mentioned the potential spillover of violence from that country. More generally, however, Botswana fear a deluge of illegal immigration from the north—from Zimbabwe, Zambia, and particularly the DRC.

Clearly, the Chobe River is regarded as a border. The behavior of both states, in mobilizing troops, in ultimately deferring to an international court for adjudication, and, most recently, in Namibia's decision to "respect" the court's finding that the disputed territory belongs to Botswana, serves to reinforce popular perceptions of sovereignty and maintain the state.

Reframing "Security" as Ecosystem Sustainability

But perhaps there is another way to frame thinking about Sedudu and its role in regional security—one that will foster longer-term stability, especially because, over time, the channel of the Chobe River will no doubt change again and Sedudu will once again become an issue in international politics. A beginning may be made in locating Sedudu within the regional ecosystem. The island has great value, in particular, to the local ecology of Chobe National Park. As a wetland, the island is home to a wide variety of flora and fauna; indeed, as stated earlier, ecologists suspect that it is probably the large hippo population residing in and around Sedudu that accounts for the fact that the channel has changed its course and depth, bringing this issue before the ICJ. At the same time, the island is a natural stopping point in the transmigration of large mammals, especially buffalo and elephant. Sedudu's value, therefore, is not unlike that of the Okavango delta. The island and the river are important components of the national park. The national park marks the confluence of the interests of numerous stakeholders in Botswana: government, small-scale farmers, tourism operators, indigenous peoples, intergovernmental organizations, and local and international NGOs—in other words, the very same groups interested in the preservation of the delta. At the same time, the Namibian interest groups are, again, small-scale farmers and cattle keepers. Given that land in the Eastern Caprivi Strip is the most fertile of Namibia's limited arable resources, it is doubtful that these farmers will make way for a superpark. However, it is not impossible to conceive of them as ultimately part of the management structure of a superpark like that emerging around Gemsbok National Park, which straddles Botswana's southwestern border with South Africa. The role of Conservation International, and that group's success in establishing a trans-Caprivi peace park, is crucial to the further extension not only of the peace park's borders, but also to the extension of regional peace.

Problems in the Chobe River region mirror those in the Okavango River region. Each, however fraught with the discourse of conflict, serves as an emergent and fungible model of conflict resolution and cooperative regional behavior: the creation of interstate institutional structures, the marshaling, however contradictory, of expert and popular forces, and open dialogue at the substate, state, and interstate levels.

Conclusion: Avoiding Clausewitz, in Search of Kant

This chapter has highlighted the centrality of the environment in regional politics in southern Africa. Clearly, there are great difficulties in achieving peace and security through environmental cooperation, particularly when the most common conceptualization of the environment is that of resources for capture, and when the dominant framework within which decision makers operate is classical realism. Difficulties are in part historical, with contemporary inequalities reflective of a pattern of human settlement based on race and the massive exploitation of the region's mineral and agricultural resources. Over time, the region's incorporation into the global capitalist economy wove small pockets of great wealth into a great fabric of grinding poverty. Environmental problems reflect these inequalities.

For each of the four examples provided (Angola, Lesotho, the DRC, and Botswana and Namibia), one could make a convincing argument that the pursuit of "state security" was in fact the defense of elite privilege. Only in the case of Botswana and Namibia does it seem possible to link security with environmental cooperation beyond the narrow agenda of "resource capture" (although even in this case the governments in both countries clearly conceive of water as power). Given the similarity of the conflicts in Botswana and Namibia to conflicts within and between other states and river basins in the region, the Okavango delta example in particular may form the basis for a wider strategy of environmental peacemaking in southern Africa. The high-political ways in which the Okavango and the Sedudu cases were framed as national security issues, however, serves to caution those wishing to read too deeply into the peacemaking possibilities of interstate environmental interdependencies.[41]

At this point it may be useful to structure conclusions and observations around the four claims made at the outset. The first was that overlapping ecological interdependencies lead toward the development of common resource management regimes. Based on the above evidence, this assertion is correct: the Highlands Water Project, OKACOM, SADC's shared watercourse systems protocol, and the SAPP are all testimony in this regard. It is not clear, however, how these management regimes contribute to regional peace. Along with Botswana and Namibia, Angola is party to OKACOM; along with the rest of SADC, the DRC is a central actor in the SAPP; Lesotho and South Africa cooperate on the Highlands Water Project. Yet Angola, the DRC, and Lesotho are also fraught with instability. Indeed, low-political agreements seem to be reached, and made functional, without any positive impact on the regional peace at all.

With regard to the second proposition, that trends toward post-Westphalian governance are leading to more sustainable and equitable development practice in the region, the evidence is equivocal. In the case of Botswana and Namibia and the Okavango River basin, the complex web of decision-making and interest-group structures suggests that civil society, in connection with wider, international interests, can help to drive the pace and direction of discourse on certain issues. The language has sometimes been bitter, but at each stage, interested parties have gone back to their constituencies to reflect on the present state of affairs. Clearly, this is much better than unilateral decision-making by authoritarian leaders. At the same time, however, it must be pointed out that the NNEWC pipeline is likely to be completed. One is given to wonder whether or not this may be just another example of state-makers, technocrats, and engineering "experts" ultimately having their way (albeit after very frank discussion with an engaged local and global community). Given that interbasin transfer schemes are likely to be the "wave of the regional future,"[42] the positive process underway in the delta may ultimately be overshadowed by those wishing to point to the outcome. Clearly, the leaders of Zimbabwe would like to see the NNEWC pipeline completed in Namibia so as to demonstrate the "viability" of their own pet project linking the city of Bulawayo to Batoka Gorge: if a project goes forward in such a delicate ecosystem as the Okavango, what does one more dam along the Zambezi matter? Until demand management becomes a central component of regional water-supply strategies, "scarcity" will be a persistent and, contrary to popular opinion, largely self-inflicted problem.

The third proposition was that emphasis on functional, often low-political issue areas binds regional state-makers and stakeholders together in such a way as to offset the divisive, negative, and potentially conflictual reality of weak formal economies. The evidence seems to suggest that state-makers are more than willing to manipulate whatever issue comes their way for political gain, irrespective of whether the issue is framed as a question of "national security" or "ecosystem sustainability." At the same time, in the SADC region, there is an overwhelming propensity to frame "security" in a state-centric, high-political, and military-privileging way. In fact, as long as resources are considered as fragmented sources of power, resource use issues will continue to drive the war in the DRC: Namibia wants the Congo's water; Zimbabwe wants mineral concessions and hence enhanced capacity to generate foreign exchange. No matter how many "protocols" SADC's states enter into, be they environmentally oriented or otherwise, it seems that in this debt-distressed and deeply divided region, elite-defined high-political issues take precedence over all others.

Lastly, with regard to the fourth proposition—that, taken together, these developments help move the region toward a peaceful and (shared) pros-perous future, as much as one might like to say that recent developments are leading toward peace and prosperity in the SADC region, the evidence at hand seems to point in a different direction. Peace and (shared) pros-perity seem to be elusive for the following reason: the global context

of neoliberalism encourages states to "go it alone," to pursue comparative advantage through highly destructive and racially exclusive economic
practices, and to therefore view land in the region as nothing more than
an exploitable (or underexploited) commodity. The global context, therefore, exacerbates the pathological tendencies of failed Westphalian state
projects and undermines even the best intentions of those leaders oriented more toward state- and nation-building projects. To build their nations in a fraught and fragmented region, these leaders must follow a soli-
tary path.

A strategy for environmental peacemaking must begin with a region-
wide approach to the management of demand for water. To this end, South
Africa's new water law, as well as the Department of Water Affairs and
Forestry's emphasis on citizen action for the elimination of exotic species,
is a simple and obvious way to start. So too are Zimbabwe's and South
Africa's development of Catchment Councils and Catchment Authorities.
These institutions are not wedded to state boundaries in decisions regarding
resource use. They are in tune more with the rhythms of ecosystems than
with the peccadilloes of state-making elites. Supply-side interventions, like
interbasin transfer schemes, can only lead to conflict between peoples and
to environmental degradation, in both the short and the long run. Unfortunately, such interventions also involve big capital and "experts" and appear
very attractive to governments in the region.[43] Reducing waste, improving
allocation among competing users, encouraging crop switching among
farmers: who would have imagined that such local, community-based activities would be at the heart of the region's foreign policy for the twenty-
first century?

Notes

1. Colin Gray, "Clausewitz Rules, OK? The Future Is the Past—with GPS," *Review of International Studies* 25, no. 5 (December 1999): 161–82.
2. Fred Halliday, "The Potentials of Enlightenment," *Review of International Studies* 25, no. 5 (December 1999): 105–26.
3. For a more detailed discussion of these and other themes in the international relations of southern Africa, see Peter Vale, Larry Swatuk, and Bertil Oden, eds., *Theory, Change, and Southern Africa's Future* (London: Palgrave, 2001).
4. Larry A. Swatuk and David R. Black, eds., *Bridging the Rift: The New South Africa in Africa* (Boulder: Westview, 1997); and Fredrik Söderbaum, "The New Regionalism in Southern Africa," *Politeia* 17, no. 3 (1998): 75–94.
5. SADC's predecessor was known as SADCC (the Southern African Development Coordination Conference). The latter was created in 1980 with nine member states:

Angola, Botswana, Lesotho, Malawi, Mozambique, Swaziland, Tanzania, Zambia, and Zimbabwe. Namibia became its tenth member when that state became independent in 1991. The coordination conference was superceded by SADC, formed by treaty in 1992. South Africa joined SADC in 1994, Mauritius in 1995, and Seychelles and the Democratic Republic of the Congo in 1997.

6. See, in particular, U. S. Agency for International Development, Regional Center for Southern Africa (RCSA), "Regional Integration through Partnership and Participation," *RCSA Strategic Plan 1997–2003* (Gaborone: RCSA, 1997).

7. For a theoretical treatment, see Hussein Solomon, "Realism and Its Critics," in Vale, Swatuk, and Oden, eds., *Theory, Change, and Southern Africa's Future,* 34–57.

8. See, for example, Peter Fabricus, "US-SA Armed Forces Strengthen Bonds," *Cape Times,* February 23, 2000.

9. "Neo-Europes" are "the zones of the earth's surface richest in photosynthetic potential [that] lie between the tropics and fifty degrees latitude north and south. There most of the food plants that do best in an eight-month growing season thrive. Within these zones the areas with rich soils that receive the greatest abundance of sunlight and, as well, the amounts of water that our staple crops require—the most important agricultural land in the world, in other words—are the central United States, California, southern Australia, New Zealand, and a wedge of Europe consisting of the southwestern half of France and the northwestern half of Iberia. All of these, with the exception of the European wedge, are within the Neo-Europes; and a lot of the rest of the Neo-European land, such as the pampa or Saskatchewan, is nearly as rich photosynthetically, and is as productive in fact, if not in theory." Alfred W. Crosby, *Ecological Imperialism: The Biological Expansion of Europe 900–1900* (Cambridge, U.K.: Cambridge University Press, 1986), 305–6.

10. Timothy M. Shaw, "Kenya and South Africa: 'Sub-imperialist' States," *Orbis* 21, no. 2 (Summer 1977): 375–94.

11. See Ronald Libby, *The Politics of Economic Power in Southern Africa* (Princeton: Princeton University Press, 1987).

12. Sam Moyo, Phil O'Keefe, and Michael Sill, *The Southern African Environment: Profiles of the SADC Countries* (London: Earthscan, 1993), 270.

13. Ironically, by recognizing these inherited "boundaries," Africans through the Organization of African Unity (OAU) reinforced the view that mountains, rivers, and lakes that for eons had determined and developed an intraregional ebb and flow of human relations were naturalized as sites of exclusion. Yet, as John Keegan points out, "large rivers, highland barriers, [and] dense forests form 'natural frontiers' with which, over time, political boundaries tend to coincide; the gaps between them are avenues along which armies on the march are forever drawn." Perhaps the OAU members are only enshrining in law one of the less admirable human tendencies, particularly under capitalism: the desire to exclude and make "mine." See John Keegan, *A History of Warfare* (Toronto: Vintage, 1994), 71.

14. The World Bank offers no data for what it considers to be small and micro states, into which category both Swaziland and Seychelles fall. World Bank, *World Development Report 2002: Building Institutions for Markets* (Oxford: Oxford University Press, 2002).

15. Manfred Bienefeld, "Structural Adjustment and the Prospects for Democracy in Southern Africa," in David B. Moore and Gerry J. Schmitz, eds., *Debating Development Discourses* (London: Macmillan, 1995).

16. World Resources Institute et al., *World Resources: A Guide to the Global Environment* (New York: Oxford University Press, 1998), 248; and *Mmegi/The Reporter,* September 25–October 1, 1998.

17. This problem is characteristic of supply-centered thinking. Recently, there has been a great deal of pressure to rethink strategies for overcoming "scarcity" by addressing demand: e.g., appropriate pricing for water, removal of invader species, recycling of water (particularly in the mining industry), and incentives for crop switching—away from high-water-using or low-return-on-investment export crops toward more appropriate, drought-resistant, and regionally profitable food crops. See Larry Swatuk, "The Environment, Sustainable Development, and Prospects for Southern African Regional Cooperation," in Swatuk and Black, eds., *Bridging the Rift.*

18. For example, in South Africa overstocking in the Karoo has resulted in the virtual desertification of this former shrub savannah. Soils in the Nama Karoo biome are easily eroded and require 5 hectares per sheep. Even more delicate is the Succulent Karoo biome, with sustainable stocking rates of 9 hectares per sheep. Serious overgrazing has led to both desertification and bush encroachment. See Phyllis Johnson and Munyaradzi Chenje, eds., *The State of Southern Africa's Environment* (Harare: Zimbabwe Publishing House in cooperation with SADC/SARDC/World Conservation Union [IUCN], 1994).

19. See Brian Davies and Jenny Day, *Vanishing Waters* (Cape Town: University of Cape Town Press, 1998).

20. Keegan, *A History of Warfare,* 73.

21. Defense spending still ties up between 5 and 16 percent of the annual budgets of SADC member states. With the resumption of war in both Angola and the DRC, many of these figures will have to be revised significantly upward.

22. Thomas F. Homer-Dixon, "Environmental Scarcities and Violent Conflict: Evidence from Cases," *International Security* 19, no. 1 (Summer 1994): 5–40.

23. Robert Jackson, *Quasi-states: Sovereignty, International Relations, and the Third World* (Cambridge, U.K.: Cambridge University Press, 1990).

24. The 1884–85 Berlin Conference instituted the basis for postcolonial state sovereignty by dividing the continent not on the basis of African history and interests, but on the basis of European needs and desires. As a consequence, the juridical boundaries of the postcolonial African state included many long-independent nations within these inherited territorial markers. This process contrasts with the European process of single, self-proclaimed "nations" fighting for, creating, and maintaining states. Hence the term "state-nations" rather than "nation-states."

25. Moyo, O'Keefe, and Sill, *The Southern African Environment,* 178.

26. On the South African case, see Val Percival and Thomas F. Homer-Dixon, "Environmental Scarcity and Violent Conflict: The Case of South Africa," *Journal of Peace Research* 35, no. 3 (May 1998): 279–98; on regional water supplies see Hussein Solomon, ed., *Sink or Swim? Water Resource Security and State Cooperation,* Institute for Defence Policy (IDP) monograph series no. 6 (Midrand: IDP, 1997).

27. For an extended discussion of violence as a consequence of poverty, see Swatuk, "The Environment, Sustainable Development."

28. The author thanks J. Zoë Wilson for providing much of the information in this section.

29. For an interesting exploration of African states' abilities to "broadcast"

power—i.e., exercise power over people and territory—see Jeffrey Herbst, *States and Power in Africa* (Princeton: Princeton University Press, 2000).

30. For a concise history of these events, see Larry A. Swatuk, "Troubled Monarchies: Democratic Struggles in Lesotho & Swaziland," *Southern Africa Report* 10, no 4 (June 1995): 30–35; and Roger Southall, "Democracy at Gunpoint? South Africa Intervenes," *Southern Africa Report* 14, no. 1 (December 1998): 12–17.

31. A "parastatal" is an entity that is partially government-owned.

32. Peter Vale and Sipho Maseko, "South Africa and the African Renaissance," *International Affairs* 74, no. 2 (April 1998): 271–87.

33. On the potential benefits of hydropower development in the DRC to regional cooperation on global warming, see Ian H. Rowlands, *Climate Change Cooperation in Southern Africa* (London: Earthscan, 1998).

34. Ken Conca, "Environmental Cooperation and International Peace," in Paul F. Diehl and Nils Petter Gleditsch, eds., *Environmental Conflict* (Boulder: Westview, 2000).

35. In Phyllis Johnson and Munyaradzi Chenje, eds., *Water in Southern Africa* (Harare: ZPH in cooperation with SADC/SARDC/IUCN, 1996), 163.

36. The Okavango River system is an endoreic system (i.e., one with terminal lakes and an interior drainage basin) whose headwaters are located in the southwestern Angolan highlands and that ends in the swamps of the Okavango delta.

37. The Okavango River basin covers an area of 570,000 km². The river itself is 1,100 km long and has an estimated mean annual runoff (measured at river mouth) of 11,650 million m³. This may be compared with the Zaire River, with its 3,800,000 km² basin, 4,700 km length, and 1,260,000 million m³ mean annual runoff; and the Zambezi, with its 1,400,000 km² basin area, 2,650 km length, and 94,000 million m³ mean annual runoff. See John Pallett, ed., *Sharing Water in Southern Africa* (Windhoek: Desert Research Foundation of Namibia, 1997); and A. I. Conley, "A Synoptic View of Water Resources in Southern Africa," in Solomon, ed., *Sink or Swim?* 17–69.

38. According to John Pallett et al., "Water abstraction from the Okavango does not pose a serious threat in itself as long as it is done with sensitivity. Ideally this should ensure that the quantity taken out is less than 1–2 percent of the total inflow, it is done at the apex, and sediment supply is not disrupted. In other words, a pipeline from the apex to where it is required is the best solution. Any dam or weir built at the apex or close upstream would stop sandy sediment entering the Swamps, and affect the dynamics of channel switching and rejuvenation. This would result in gradual destruction of the system." Pallett, ed., *Sharing Water in Southern Africa,* 30–31.

39. Moyo, O'Keefe, and Sill, *The Southern African Environment,* 190.

40. Ibid., 169.

41. It should be noted that I have deliberately excluded the case of the Zambezi River basin, partly because I have treated it quite extensively elsewhere. See Larry A. Swatuk, "Power and Water: The Coming Order in Southern Africa," Southern African Perspectives working paper no. 58 (Bellville: University of the Western Cape, Center for Southern African Studies, 1996).

42. See Swatuk, "The Environment, Sustainable Development."

43. Lori Pottinger, "Damn the Lesotho Dam," *Southern Africa Report* 13, no. 2 (July 1998): 21–25.

6

Beyond Reciprocity: Governance and Cooperation around the Caspian Sea

Douglas W. Blum

The setting for a recent James Bond movie, the Caspian Sea has achieved notoriety as a site of considerable oil and gas wealth and Byzantine political intrigue. Yet despite the well-publicized disputes in the realm of high politics, there has also been significant movement toward regional cooperation in environmental protection. This chapter asks how and why such cooperation is possible amid the geopolitical clashes and high-stakes oil gambles that preoccupy the media. In addition, it considers the larger potential importance of such cooperation, even beyond its immediate environmental objectives. Are there likely to be any substantial spillover effects for regional peace and normalization? Considering this question involves an examination of the relationships among Caspian countries and the factors that influence them, as well as the functional and political aspects of the nascent cooperative arrangement.

The international organization that has been created to foster ecological security is the Caspian Environmental Program (CEP). Because of its central role in the emergence of this cooperative process, this chapter discusses the CEP at some length. To foreshadow one of the key findings, the CEP embodies an innovative process of "governance" that reflects both the opportunities of interdependence as well as the challenges it poses to state sovereignty. Bringing together central and municipal governments, international organizations, transnational scientific knowledge, and local nongovernmental organizations (NGOs), the CEP introduces elements of what may be called "post-Westphalian" organization.[1] What difference, if any, does this make? Without making any assertion that the CEP has already achieved concrete objectives, the point here is to explore its prospects for success and to identify the factors on which such success is contingent, as well as the implications of this process for the member states involved. In particular, to the extent that environmental cooperation becomes consolidated and well institutionalized, what societal, normative, or other domestic political changes are likely to follow?

This chapter begins with a brief overview of regional environmental problems and moves on to examine regional dynamics at the interstate level, including political-economic factors and collective identities that have obstructed cooperation in general. It then focuses on the CEP itself and discusses its functional components and processes as well as their suitability for overcoming environmental problems and fostering political normalization. The final section of the chapter evaluates the normative and social-organizational dimensions of the CEP and considers the implications of such a normative reorientation for domestic and transnational politics, as well as for cooperative action. In the end, the form of cooperation and the social channels through which it operates are likely to help shape its effects in both functional and political terms.

Caspian Environmental Problems

The Caspian Sea is beset by fisheries depletion, water pollution, and a dramatic fluctuation in sea level.[2] These problems in turn are linked to numerous additional causes and effects, since changes in sea level combined with pollution have undermined the viability of coastal wetlands and threatened the entire regional ecosystem.

The depletion of fish stocks is largely a result of the fall of the Soviet Union. During the Soviet period, quotas for catch of commercial stocks were established and laws against poaching were enforced. However, with the emergence of four new states and the decline of any central regulatory agency, overfishing and poaching have become rampant. The most well-known problem is the threat to the sturgeon population, since the Caspian has historically been responsible for 90 percent of the world's sturgeon catch as well as the highest quality caviar.[3]

Caspian fisheries are further depleted by pollution.[4] To be sure, pollution is not a new problem. The Caspian's water quality has been in decline for many years, largely due to the Soviet command system with its output quotas and lax environmental enforcement.[5] The leading source of water pollution is waste discharge from rivers (especially the Volga) and from factories and towns on the coast.[6] The increase in oil and gas extraction and transportation has already had a discernible effect on pollution, which may further damage fisheries and migratory bird breeding sites unless future drilling operations are carefully controlled.[7] Besides chemical and sewage pollution, there is also a hazard of radioactive exposure, since the northern Caspian borders several sites where underground nuclear detonations were once conducted.[8] Thus, in addition to its detrimental effect on fisheries, water pollution jeopardizes tourism and poses a risk of serious public health problems.

The fluctuation in sea level, too, began under Soviet rule. After falling sharply (by as much as 4 meters) from 1930 to 1977, the Caspian Sea rose nearly 2.5 meters between 1978 and 1996. Since then, it has begun to sub-side—apparently the latest cycle in what remains a poorly understood pro-cess of fluctuation.[9] Unfortunately, Soviet bureaucrats did not factor such changes in water level into their planning, as a result of which many indus-trial facilities and entire communities have been inundated, while many thousands of hectares of cultivated land have been flooded and salinated. Even if the water continues to recede, the damage already done is enormous.

In addition to the above-mentioned problems of water pollution, fish-eries exhaustion, and sea-level change, the Caspian region also suffers from serious desertification in certain areas, especially in Turkmenistan and Kalmykia. Moreover, fauna native to the sea and its famous coastal refuges—such as the Caspian seal and numerous rare species of birds—face the threat of catastrophic depletion.[10] Many environmentalists empha-size the potential threat from poorly managed oil and gas development.[11]

The scope and interconnectedness of Caspian environmental problems are such that they cannot be addressed adequately by unilateral state efforts. Fish stocks and clean water are classic "common pool" resources, the management of which requires organized collective action. Other environ-mental problems in the Caspian basin are dispersed and wide-ranging; water pollution arises from point-specific effluents as well as major estu-aries, and from microeconomic actors as well as multinational oil com-panies. Without concerted efforts, the costs of managing such problems will be asymmetrically distributed and responses are likely to be only marginally effective. The same is true of flood control: unless they are well-coordinated, management efforts undertaken in a single area will be

Table 6.1

Human Development Statistics for the Caspian Region

Country	HDI rank[1]	Life expectancy at birth (years) (1998)	GDP per capita (PPP, U.S.$)[1] (1998)
Russia	62	66.7	6,460
Kazakhstan	73	67.9	4,378
Azerbaijan	90	70.1	2,175
Iran	97	69.5	5,121
Turkmenistan	100	65.7	2,550

Source: United Nations Development Programme (UNDP), *Human Development Report 2000* (New York: Oxford University Press, 2000).
[1]For an explanation of the Human Development Index (HDI) and purchasing power parity (PPP), see the notes to Table 2.1 on page 26.

ineffective overall. Indeed, flood barriers raised in one area may cause even higher water levels in other parts of the closed basin.[12]

It should be stressed that the existence of ecological problems is more than an abstract or transcendent concern. The combination of extensive water pollution, fisheries depletion, and sea-level fluctuation poses a social and political threat throughout the basin. Serious environmental degradation affects thousands of stable livelihoods and could conceivably jeopardize the ability of communities and state agencies to function cohesively. Strains on social stability in this underdeveloped, polyglot ethnic region are easy to imagine, as well (see Table 6.1). Indeed, because of the prevailing pattern of ownership and income distribution, the current practices of energy extraction, transportation, and production lead to wider income disparities and ethnic cleavages.[13] This is not to imply that environmental management is necessarily unproblematic. Conceivably, environmental restrictions might create tension over access to resources that have traditionally been linked to particular ethnic groups.[14] But in the context of already-heightened interstate tensions, the increasing gravity of environmental problems (and the attribution of blame) may help to create a politically combustible climate. If severe enough, environmentally aggravated economic dislocation could lead to migration, institutional decay, and overt political instability.[15]

High Politics and Environmental Cooperation

Since the fall of the Soviet Union, the Caspian's legal regime has become a contentious issue among the states that border it.[16] This squabbling was magnified enormously by the discovery of potentially large quantities of undersea oil and gas, culminating in the fall of 1994 when Azerbaijan asserted ownership over three oil deposits and granted a development license to a consortium of private oil companies. The ensuing scramble for resources generated fierce political competition, spawning an equally intense, and often acrimonious, struggle for control of pipelines leading from the Caspian to external markets.[17] One immediate result was that ecological issues became intertwined with complex considerations of legal ownership of natural resources. This in turn involved a close intermingling of questions about environmental responsibility, state liability, and rights to resource extraction. The discovery of potentially massive deposits in the northern Caspian in the summer of 2000 reintensified these rivalries, which had declined somewhat over the previous two years because more gas than oil tended to be found in the drilling conducted off the shores of Azerbaijan.

During the period leading up to 1998, resource competition precluded virtually any form of regional collaboration. To some extent this was a product of the political power of elites with direct interests in energy exploitation within all the littoral states. A far more widespread factor, however, was the prevalence of traditional, geopolitical perspectives. Such perspectives, in turn, were partially linked to heightened sensitivity about national autonomy. These concerns were particularly pronounced on the part of the former Soviet states of Azerbaijan, Turkmenistan, and Kazakhstan, where political elites emphasized the continued existence of a military threat from Russia.[18] Success in exploiting oil and gas deposits was seen as a means of gaining sufficient wealth (and powerful political alliances) to guarantee independence. For example, speaking about gaining access to pipeline routes to the West, the director of the Kazakhstan Institute of Strategic Studies, Ermukahmet Ertsybaev stated, "Of course Russia will not agree with this project, and will try to block it. It would have a huge political impact. We have been isolated from the world system, and this would be our link to Europe. It would make our independence real, and also the independence of Azerbaijan and Georgia."[19]

Such attitudes reinforced, and were reinforced by, the evolution of post-Soviet identities in Russia. For a significant segment of the Russian political elite, the articulation of national identity involved a self-image as imperial hegemon, and a demand for international recognition as a "great power."[20] In Iran as well, competitive pragmatism increasingly replaced revolutionary messianism as the cornerstone of foreign policy.[21] Whatever their source, the ubiquity of relative-gains interests and realpolitik views quickly led to an impasse in negotiations on legal ownership, as several of the littoral states began to claim sovereign property rights to subsoil hydrocarbons located within "their" sector of the sea.

The emergence and collision of these identities created enormous obstacles for regional environmental collaboration. After all, environmental protection requires a substantial shift in the allocation of resources and thereby imposes opportunity costs. To the extent that current consumption is valued over long-term conservation, national leaderships are likely to be less willing to implement stringent environmental regulations. Thus the degree of political and resource competition is important, inasmuch as it accentuates the prevailing short time horizon. For these reasons the "geopoliticization" of the Caspian Sea, including the involvement of the United States, other Western states, and multinational corporations, constitutes an important obstacle to shared governance.[22]

Despite the existence of political strains, however, the leadership of every Caspian state recognized the desirability of environmental cooperation, at least in the abstract. Indeed, this recognition predated the frantic competition for oil and gas. The first formal effort can be traced to Tehran in October 1992, when the Iranian government convened a gathering of senior officials from all of the bordering countries and proposed creating a regional organization for this purpose. The reason for this Iranian initiative appears to have been primarily environmental, inasmuch as national economic interests (fisheries, coastal villages) were vulnerable to environmental dislocation following the collapse of the Soviet-Iranian legal regime in the Caspian.[23] Here, then, is an example of environmental cooperation emerging at the level of high politics.

Subsequently a number of similar meetings took place, attended by representatives of the ministries of environment and national task forces on Caspian problems.[24] Although these early meetings produced a number of general framework proposals, no significant implementation occurred. More often than not, discussion of environmental cooperation was overshadowed by haggling about access to oil wealth. Moreover, the Russian position in support of stringent environmental regulations was widely— and correctly—regarded as a subterfuge designed to prevent the former Soviet states from extracting oil and gas from "their" sectors.[25] Thus, early on, environmental cooperation fell hostage to tensions in high politics, even though official government initiatives were responsible for launching the process.[26]

The Caspian Environmental Program

The logjam in regional environmental cooperation began to loosen with the advent of the Caspian Environmental Program in 1995. The CEP initially emerged through the efforts of the United Nations Environment Programme (UNEP), the UN Development Programme (UNDP), and the World Bank, although it was subsequently joined by the European Union's Technical Assistance for the Commonwealth of Independent States (TACIS) program and received start-up funding from the Japanese government. With incipient but rather passive governmental support, environmental activists within some of the leading international organizations began pushing for development of a Caspian program modeled on the Black Sea Environmental Program and similar efforts elsewhere.[27] In late 1998, the CEP was brought under the auspices of the Global Environment Facility, and at various stages the UN Educational Scientific, and Cultural Organization, the International Maritime Organization, and the International Atomic Energy Agency have also participated, as well as numerous NGOs under the coordination of the U.S.-based Initiative for Social Action and Renewal in Eurasia (ISAR). Building on previous attempts to gain environmental agreement, the CEP systematically links cooperation across a wide range of issue areas, including management of sea-level fluctuation, pollution prevention and monitoring, and biodiversity protection.[28] In short, it is a classic cooperative regime.[29]

The CEP includes three main organizational components. First, each of the five states is assigned primary responsibility for establishing a "Thematic Center" to oversee specific areas of functional cooperation. The states have been assigned the following Thematic Centers: Azerbaijan, pollution control and data management; Iran, intercoastal zone management, emergency response, and pollution monitoring; Kazakhstan, biodiversity and water-level fluctuations; Russia, fisheries and bioresources and legal and regulatory issues; Turkmenistan, desertification, sustainable human development, and health.[30] Second, actions and communications among the member states are managed by a Program Coordinating Unit (PCU), whose location is to rotate among them. The PCU was established in Baku in 1998. The location of the PCU as well as specific country assignments were negotiated by country representatives during an all-night session at a CEP meeting in Ramsar, Iran, in May 1998.[31] The third formal organization within the CEP is the Regional Steering Committee, which is charged with setting the long-term agenda and making strategic decisions. The Steering Committee, which includes both governmental and non-governmental representation, is "comprised of Deputy Ministers of Environment or equivalent rank individuals from the Caspian littoral countries and representatives of international organizations, bilateral programs, and other organizations that actively support the CEP."[32]

The CEP incorporates a number of features designed to foster stable cooperation and that embody institutional lessons taken from model programs in the Mediterranean, Black, and Baltic sea regions. Through the workings of the central PCU and by holding regular meetings of the Steering Committee, the CEP has been crafted to include mechanisms for ensuring transparency and accountability. Although taking place at different levels, both of these institutions address concerns about relative gains and free riding and therefore mitigate collective action problems.[33]

At the same time, the program is essentially process-oriented rather than focusing exclusively on targets and deadlines. The fact that primary responsibility for providing mutually desired outcomes is apportioned among the states helps ensure that all parties perceive ongoing contributions from others, even though many of the concrete benefits will not materialize fully for a number of years. Partly for this reason, there is a conscious effort to lengthen the expected time horizon for cooperation. As frequently stated in the program's literature and emphasized by its senior officials, the CEP envisions a 15–20 year period for the achievement of its ultimate goals. (In part, too, this appears to reflect an internal, bureaucratic interest in avoiding premature program evaluation.) In these ways it institutionalizes diffuse reciprocity, which itself contributes to the durability of cooperative arrangements.[34] Finally, as will be discussed below, to facilitate implementation and long-term support for the program, the CEP attaches considerable priority to capacity building and to increasing public awareness of ecological security matters.[35]

Largely due to these design features as well as the generally rising level of concern over the Caspian environment, the CEP managed to confound pessimistic predictions by gradually gathering momentum and achieving operational launch in 1998. The governments of the participating states have issued numerous statements endorsing the program and have generally complied with its agenda in the early stages of implementation.[36] Therefore, at least for the time being, the likelihood of official obstruction seems low.

It is noteworthy that the CEP was able to begin operation without prior ratification of an Environmental Framework Convention. Although counterintuitive, this fact is hardly surprising: drafted by UNEP, the Framework Convention has received opposition from various government quarters (especially Azerbaijan and Iran) due to the legal ownership dispute and the fear that certain environmental obligations might also imply juridical constraints on energy extraction and transportation.[37] Indeed, the lack of a clear legal status may actually promote the CEP's ability to begin functioning, since given the ambiguity over ownership all resources must tentatively be regarded as shared. This situation magnifies each state's interest in cooperating with others, which might not be the case if the water column were divided into exclusive sectors. For these reasons, although the littoral states have endorsed the general objectives and process of the program, they have shied away from formally approving it.[38] Nevertheless, the ability of the CEP to mobilize de facto state recognition, and to complete an ambitious Transboundary Diagnostic Analysis in spite of such reservations, testifies to the states' shared interests in environmental protection and the potential separability of environmental and political issues in regional diplomacy.[39] Thus, although it will not be possible to evaluate the success of the program for at least several more years, the early prospects are encouraging.

In addition to political resistance to fully embracing the CEP, another barrier to the program's implementation is state economic weakness and a general reluctance to assume new loan burdens, even at relatively concessionary rates. Yet without taking on additional loans there is likely to remain a crippling shortage of capacity on the part of the post-Soviet states, including endemic political organizational weakness, lack of enforcement ability, and dilapidated infrastructure. These observations again raise the question of how great a priority will be attached to the resolution of environmental problems.[40]

Environmental and Political Cooperation: Causal Arrows

What potential significance does the CEP have for interstate security? On one level this question involves the program's success in addressing sources of economic and social dislocation. On another level, however, it engages the possibility that environmental regimes can strengthen regional

political stability by providing an institutional framework for the negotiation of other shared interests. Although not explicitly suggested in the
CEP's literature, this idea figures prominently in the hopes of its archi-
tects.[41] Indeed, there is a plausible theoretical foundation for suggesting
that environmental cooperation can promote peace.[42] This suggestion is
predicated on the construction of a robust regime encompassing extensive
and mutually beneficial collaboration; this requirement in turn implies the
concerted political efforts of state actors, in contrast to early notions of low-
level, apolitical functional cooperation leading to political integration.[43]
Ostensibly, by effectively addressing common concerns and providing significant benefits, such a regime may become a vehicle for diffuse and self-
sustaining reciprocity. Along the way, it might also help to stabilize expectations, generate deeper shared interests, and lead to still broader functional
and political cooperation. In sum, despite the existence of interstate tensions, joint action to resolve shared environmental problems may help
catalyze a process of political normalization.[44]

    As it stands, the CEP offers a potentially workable framework for
developing further agreements to promote economic interdependence
among the Caspian littoral countries. For fishing, this would involve establishing a monitoring and quota system in which the portion allotted to each
country would be determined by a set of variables reflecting each state's
contribution to total stocks.[45] A host of measures is being planned for
combating water pollution and providing rational water management: zon-
ing coastal areas for various uses; establishing monitoring centers, criteria,
and baseline values; identifying point-specific sources of pollution and
establishing viable controls; determining optimal riparian effluent regimes
for the major estuaries; creating oil-spill prevention and rapid-response
strategies; setting standards for ballast and wastewater discharge from
ships; and regulating offshore drilling activities.[46] A key sticking point
might be restrictions on certain oil and gas exploitation, such as specific
technical requirements for drilling and transportation in the northern "shallow water" zone, or even prohibition in certain areas. Efforts to address sea-
level change will focus on monitoring (and understanding) the process of
fluctuation as well as coordinating construction of drainage, retaining
walls, sluices, and other forms of intervention. Achieving these goals
would provide substantial benefits for the littoral states.

    The pacifying potential of the CEP thus appears considerable in the
abstract, especially if it were expanded to include an emphasis on sustainable development. Although this term has become rather cliché, it

does serve to capture the critical compromise between environmental conservation, social welfare, and elite political power grounded in resource extraction. Particularly in the Caspian Sea, with its recent history of intense competition for oil and gas, the prospects for achieving a viable regime hinge on the ability to accommodate extractive interests by not banning oil and gas recovery outright. Accordingly, the overarching Caspian institutional framework would seek to balance private ownership with a system of public regulatory measures linked to social and environmental concerns.[47]

Ideally, similar guidelines might be formulated to address pipeline construction[48] and might be supplemented by agreements for navigation and energy transportation, including environmental standards, guarantees of open access to shipping lanes and port facilities, and uniform transit fees. Conceivably, the legal impasse might be resolved on the basis of such guarantees. The 1998 agreement on stratified ownership between Russia and Kazakhstan (involving individual state ownership of seabed resources and shared ownership of fishing stocks in the central portion of the sea) could provide a solid legal foundation for such a sustainable development regime, if it were ultimately accepted by Azerbaijan, Iran, and Turkmenistan. As of this writing, however, the three states in question continue to reject the Russia-Kazakhstan agreement. Azerbaijan (and occasionally Turkmenistan) resists the notion of joint ownership of the water column, since this might inhibit plans for laying pipelines on the sea floor, while Iran demands either condominium ownership or equal, 20 percent ownership shares of the seabed.

Yet regardless of legal ratification, there are rational grounds for enlarging these agreements to encompass a wide range of other areas. One such area might be military cooperation for limited security objectives. Although joint declarations in favor of demilitarizing the Caspian Sea have been made repeatedly since 1992, in practice states have moved in the opposite direction. In January 2001, the Russian naval fleet pointedly engaged in firing exercises while Russian president Vladimir Putin was meeting with Azerbaijan's president, Heydar Aliev, in Baku, and in July 2001 Iranian gunboats confronted Azeri research vessels in what Azerbaijan considered its own sector.[49] Other targets for expanding cooperation could include infrastructure development and coordinated customs arrangements, including communications and freight transportation through the Eurasian corridor. Ultimately, this might include the emergence of a regional common market in energy and other trade in goods and services.[50] Such a regime would bolster state interdependence by combining the functional issue linkages already envisioned in the CEP with mutually beneficial commercial regulations.[51]

Obviously, the ability to realize broader cooperative outcomes depends on attaining narrower—yet still quite ambitious—CEP goals. Yet while ambitious, the possibility of expanding interdependence is potentially realistic, depending on the degree to which successful environmental collaboration serves important state interests and creates cooperative expectations and norms. Cumulatively, through institutionalized changes in interactions and roles, this process might be expected to involve far-reaching shifts in collective identity. To the extent that such transformations occur, the context of strategic calculations may be fundamentally altered.

High Politics, Low Politics, and Environmental Governance

Thus, in the abstract, it would appear that the political gains from issue linkage within a broad and reciprocal regime framework could be substantial. Over the long term, however, the prospects for environmental cooperation's serving as a catalyst for regional political stabilization depend in part on the conceptual and institutional arrangement involved. For reasons already discussed, national strategic and domestic political agendas may obstruct regional stabilization as long as environmental collaboration is conducted exclusively at the official, ministerial level. This is an important obstacle that must be addressed both in regime design and in political practice. Doing so effectively will require astute leadership, but it will also mean establishing an independent institutional framework within which rational goals and standards can be identified in an apolitical manner.

Yet even this is not enough. Beyond the issue of institutional design, a larger problem looms: fostering the political conditions within which the purposes of the regime can be realized. Even if environmental sustainability is considered narrowly in terms of criteria for reducing emissions or maintaining wildlife populations, interstate collaboration to achieve such targets is an essential element of success. State capacity must be harnessed while at the same time nonstate actors must be included, in order to influence the process of setting goals and overseeing compliance. A broader perspective on environmental sustainability, however, would address prevailing norms, organized political interests, and institutional channels for articulating and negotiating policy options. A potentially transformative approach would therefore address the political requirements for altering such norms and recasting the relevant political process.

Indeed, a close examination of the CEP reveals precisely such an approach, while at the same time the program makes a number of major concessions to prevailing norms and relations of power. Thus, on the one hand, the CEP is traditionally state-centric in its inclusion of central government officials and its reliance on national agencies for follow-on funding. On the other hand, it envisions a complex skein of overlapping participation by scientists, local governments and communities, and NGOs operating above and below the state level. The following sections will explore these points in greater detail.

Scientific Knowledge

Scientists and scientific knowledge play a central role within the CEP in challenging dominant identities and their associated patterns of realpolitik state relations. "Objective," "factual" knowledge of cause-and-effect relationships is essential for reframing issues and relevant trade-offs, and for constraining the acceptability of conceivable policy options. To the extent that means-ends relationships are recognized to involve costly trade-offs between wealth and social well-being, elite interests may undergo substantial adjustment.[52] Clearly this process is not devoid of politics, both within the scientific field and at the nexus of science and policymaking.[53] Still, negotiation of the meanings attached to factual knowledge may yield changes in outcomes, particularly when such knowledge is grounded in strong decision-making organizations.[54]

The Thematic Centers of the CEP are to be staffed mainly by scientists from the Caspian states, assisted by international scientific advisers. Working contacts are strongly encouraged among the centers, as well as between the centers and the regional scientific community.[55] The CEP has also convened a Bio-Resources Network composed of regional and international environmental experts, to serve as an ad hoc source of knowledge production and policy advice.[56] From the start, then, there has been a conscious effort to build a transnational epistemic community, both in order to influence decision-making and as an aid to functional interdependence.

Civil Society, Transnational Linkages, and International Organizations

The framers of the CEP recognize that scientific knowledge alone may be inadequate for shifting state preferences, especially in the face of organized economic and political pressure to promote an extractive agenda. For this reason, the regime reveals an effort to reshape political interests by shifting the nature of constituency involvement and, ultimately, recasting state-society relations. This involves systematic efforts to assist in the creation of civil society by networking with local environmental NGOs, based on the conviction that if social actors with an inherent stake in environmental outcomes are politically mobilized, the likelihood is increased that conservationist policies will be consistently implemented. Even without formal channels of access, the ability of mobilized social actors to exert political pressure and require accountability is likely to enhance the compliance of local officials.[57]

The inclusion of NGOs within the CEP process is thus geared to the creation of broad-based civic polities, as one necessary support for a viable regime. According to the CEP's "Concept Paper," "the CEP will include actions by governments, the private sector, and civil organizations in the region. One objective of this approach is to create transnational networks and public-private partnerships, to take actions in the stakeholders' mutual interest and to enhance the sea's sustainable development and protection."[58] In addition, "the CEP anticipates broad-based participation by the general public, private sector associations (especially oil and gas companies), academic and research institutions, non-governmental organizations and local community groups. The Program will identify key stakeholders, particularly effective NGOs, bring them together to strategize and discuss common issues, link them together for the enhanced exchange of information and strategies, and involve them in the [policy] formulation processes done on country-specific and regional bases."[59]

The stirrings of civic activism and voluntary social participation are evident in several of the Caspian states, most notably Russia, Kazakhstan, and Azerbaijan. Nevertheless, the predominant political actors in this region remain more or less centralized nation-states. Central political elites and state agencies dominate the negotiation process, independent media outlets remain scarce, and few alternative mechanisms exist for engaging the public arena. To some extent this lack of pre-existing, robust civil society limits the ability of the CEP (or any other cooperative mechanism) to exert political leverage on the Caspian states and thereby to promote effective environmental governance, at least at present. Moreover, the massive economic and social dislocation in the former Soviet region has distracted popular attention from broad-ranging and seemingly less immediate issues such as environmental protection.

In the absence of a powerful, indigenous social impetus for environmental governance on the part of the Caspian countries, international organizations have played a key role in initiating and guiding the cooperative process. Thus, while the original idea of constructing a regional environmental regime emerged at the state level and was explored in multilateral negotiations, the specific form and character of the CEP is a product of influential international organizations and key actors within them: UNEP, UNDP, the World Bank, and TACIS.[60] As some analysts would contend, this fact in itself may exert conceptual and organizational constraints on the nature of governance and the purposes it serves by reproducing state agency as a hegemonic form, or as the essential organizing principle within which the regime is embedded.[61]

Yet despite the nation-state membership of the organizations and their allegiance to the liberal world order, the CEP is explicitly and self-consciously generative of overlapping levels of governance, including state, substate, nonstate, and various international actors. As already mentioned, local and even international NGOs were intended to be included in research, training, and public education, and it was hoped they would act as watchdogs to pressure governments to execute agreements and follow the laws. Rhetoric aside, this objective is in fact being realized in the CEP's outreach activities. This has involved considerable negotiation with NGOs, many of which were suspicious initially of the CEP and resentful of not being included earlier and more extensively. The PCU in Baku has subsequently been involved in making contacts with NGOs throughout the Caspian basin, while members of the Thematic Centers have established working relations with local scientists and green activists in Russia, Kazakhstan, and Azerbaijan.[62]

To repeat an earlier point, this focus on civil society represents only one element of the CEP's philosophy. Partly because of the lack of a robust civil society in the region, the CEP has been designed and promoted as a program of the littoral states themselves rather than of its international organizational sponsors, even though in reality the latter attempt to exert a formative influence on the substance and process of negotiations. The question remains whether this international organizational component of the CEP can help substitute for civil society in the short run, while simultaneously helping to galvanize its emergence across the region. Key questions are whether environmentalists within and outside the formal organizational umbrella actually will be committed to activism,[63] and whether they can be linked in such a way as to exert significant influence on the discussion and resolution of problems.[64] It also remains to be seen whether the CEP can facilitate the large-scale emergence and participation of substate actors, especially local governments with direct interests in the Caspian environment, in the ongoing process of environmental governance. Finally, it is an open question whether this process can dovetail with—and be buttressed by—the larger development of transnational civil society.

Norms

In addition to its efforts to foster national and transnational civil society, the CEP also seeks to mobilize social pressure for environmental sustainability by fostering new constitutive norms relevant to ecological health. Interestingly, such an approach appears to reflect the emergence of an epistemic community that extends into the leading international organizations of the CEP. In addition to holding similar views about the "lessons" of previous development efforts led by international organizations, members of this transnational community share a conviction regarding the value of public activism based on the principle of sustainability.[65] Consequently, the normative component emphasizes civic education and empowerment. Moreover, this normative discourse is geared toward establishing an institutionalized process whereby "stakeholders" share responsibility for concrete policies related to sustainable development at the community, national, and regional levels.[66] Although not articulated as such, the construction and presentation of such constitutive norms is intended to encourage broad shifts in collective identity.[67]

Thus, a central norm pertains to the integrity and uniqueness of the regional ecosystem. This norm is constitutive in that it links identity with both environmental quality and place.[68] Furthermore, inasmuch as environmental concerns are inherently transboundary in nature, such a reorientation of social and political affinity challenges the territorial aspect of state sovereignty. Instead, the transcendent, intertwined character of the Caspian ecology ostensibly creates common interests among states and regional inhabitants alike. It is on this subtle level of change in values and social orientations that the CEP pins much of its hope.

Interestingly, leading figures in central government have voiced similar ideas on a number of occasions. According to the Iranian deputy minister of foreign affairs, Sadek Kharrazi, for example, the Caspian states must find a mechanism to preserve this "unique body of water" for succeeding generations.[69] In making such statements, elites have often borrowed ideas from the expert and activist communities and have used them for self-serving reasons. As Jeff Checkel's work shows, however, the instrumental manipulation of normative ideas may become politically constraining if the behaviors associated with those norms are well institutionalized.[70] Of course, to the extent that such ideas are internalized, they provide a powerful legitimizing basis for policy advocacy and subsequent action.

Sovereignty

In its boldest form, the rise of post-Westphalianism challenges the central institution of sovereignty.[71] As conceptualized here, the establishment of post-Westphalian modes of governance may have a national as well as an international dimension. The former consists of promoting civil society and creating autonomous forums for exploring and enacting collective decisions, as well as promoting more interpenetrated state-society relations. The latter, international component involves building linkages that transcend the territorial and bureaucratic confines of the state, by anchoring the processes of information gathering, evaluation, and decision-making in NGOs, epistemic communities, and transnational civil society.[72] Arguably, in polities that are highly interpenetrated by transnational civil-society relations and enmeshed in a web of transnational norms, the state's ability to fashion and execute decisions in an autonomous manner is significantly limited, and the very foundation within which sovereignty evolved may be eclipsed.[73] For the same reasons, the state's claim to the exclusive loyalty of its citizens is also cast into doubt.[74] From this standpoint, by reaching into the domestic realm in order to foster civil society and creating institutional connections across sovereign political boundaries, the CEP does introduce post-Westphalian elements.

And yet a more subtle analysis suggests that the CEP's implications for sovereignty are actually various and highly contested. After all, as Karen

Litfin has argued, the question raised by shared governance is one of not so much replacing but rather reconfiguring sovereignty in terms of its constituent parts: legitimacy, control, and decision-making autonomy.[75] Accordingly, the state's lack of independent control in a given issue area may lead it to seek compensatory joint control via cooperative measures, even while it relinquishes a measure of autonomy in the bargain. Seen in this light, the CEP does not require states to sacrifice control but rather enhances their ability to realize preferred outcomes for pollution control, biodiversity, fisheries regulation, or sea-level management. Nor does the evident propensity of the CEP to foster international linkages necessarily entail state decline. After all, NGOs may gain leverage, standing, and access to information through their inclusion in cooperative arrangements, but may also be co-opted by the state in the latter's quest for control.[76] In short, different constituencies envision markedly divergent roles for NGOs.

In addition, it should be remembered that the key functional agency within the CEP remains to a large extent the central government of each state, even if none is capable of securing its desired environmental outcomes alone. Conceivably, each state's legitimacy may be enhanced by its provision—albeit shared—of an otherwise elusive goal. In any case, the formal channels of cooperation within the CEP run through the highest echelons of state power. The pivotal role of the Steering Committee reflects the insistence on the part of the catalyzing international organizations that high-level ministries—including state agencies responsible for finance and foreign affairs—be systematically brought into the decision-making process.[77] This commitment is also expressed in the designation of "national focal points," which are nominated by each state's environment ministry and are singly "responsible for the execution of national participation" in the program.[78] Even though in the post-Soviet context such ministry-based delegation of responsibility is largely a fiction devoid of any accountability, the gesture does reveal the continued centrality of state power.

Thus, in many respects, the CEP resembles "transgovernmentalism," characterized by the disaggregation of previously closed state agencies and their international cooperation. Although this process may be suffused by a network of nonstate actors and transnational society contacts, it remains solidly bound to the state.[79] Together, these observations suggest that the CEP may in fact serve to consolidate state control—a key element of sovereignty—while diminishing state autonomy and challenging its traditional sources of legitimacy. In short, at this point in its evolution, the post-Westphalian effects of the CEP should not be overstated.

Conclusion

As the oil frenzy subsides and the lurching post-Soviet transition recedes, more prosaic matters are beginning to come into focus. As leaders seek to mitigate social upheaval, problems of stability, efficiency, and equity increasingly tend to replace grand state-building projects. Moreover, as economic globalization encroaches, the costs of balkanization increase. In short, national and international trends are likely to pull societies around the Caspian rim into close interaction and reframe the key issues of politics. Already, one such trend is evident in response to the pressure of shared environmental decay. As this early cooperative process continues, well-designed functional institutions and social networks will discover opportunities to mobilize political support, which may extend ultimately to deepening cooperation in political and economic domains. At least this is the vision of environmental activists in the region, in spite of the widespread tendency to view Caspian developments as marking the return of a geopolitical "Great Game."

Remarkably, this cooperative vision appears at least as plausible as any imagined by diplomats and realpolitik analysts. While allegedly "pragmatic" observers have focused on official disputes, drilling contracts, and pipeline schemes, many of which have dragged on without resolution for years, a plan for sweeping environmental cooperation has quietly swung into action. Addressed separately from legal ownership issues, the CEP is an encouraging example of the ability to negotiate a discrete body of shared problems, thereby allowing states to pursue environmental cooperation despite the existence of other contentious issues. Its emergence highlights the extent to which the issues involved—environmental degradation and sustainable growth—are significant interests of all of the countries involved. This fact, in turn, suggests that the CEP could indeed serve as a vehicle for identifying additional interests and promoting regional stabilization, at least over the long haul.

Even aside from the prospects for such a cooperative outcome, however, the CEP has potentially profound political significance for the region. By strengthening national and transnational civil society and increasing its access to channels of political influence, the CEP represents an

innovative combination of traditional statist and post-Westphalian arrangements, replete with new norms for public policy and collective behavior.
Participation by broader domestic and international constituencies is
directly relevant to the prospects of regime effectiveness. Yet it is also
fundamental to the development of open, democratic institutions. Thus at
the national and international levels alike, the process of environmental
collaboration may ultimately have far-reaching effects on governance and
political stability.

To sound a cautionary note, however, the program's transformative potential will unfold only gradually, if at all. The CEP is bedeviled by its own
problems, including a lack of clear organizational structure and administrative responsibility (including coordination among individual Thematic
Centers), deficiencies in leadership (leading to burgeoning resentment), and
serious questions about the adequacy of current and prospective funding.
Another problem is a lingering mentality of realpolitik, paranoia, and related competitive efforts to control information. Here, the key fears have
concerned Russia, especially regarding its alleged or anticipated imperialism, lack of openness, and imposition of preferences.[80] More general fears
of outside (presumably American) interference have also been displayed.[81]
Furthermore, the littoral states retain the ability to undermine the program
by withholding support, avoiding legal accountability, or seeking to co-opt
the platform of sustainable development for their own competitive interests.[82] This factor underlines the point that the advent of the CEP has not
suddenly altered the prevailing identities or institutional processes.

For the foreseeable future, then, debates over relevant questions of legitimacy will be influenced by prevailing ideological orientations at the national level, and by their articulation on the part of groups attempting to
form winning coalitions astride the state. Reconfigurations of sovereignty
are intertwined with reconstructions of the international political setting. As
long as the Caspian Sea is the site of unmanaged oil and gas competition,
and therefore a source of interstate tensions, cooperative norms and "bio-
regional" identities are likely to remain tenuous.

While this process unfolds, then, nation-states will hold sway as the
legitimate source of rule and resource distribution. Nor will the CEP do
much to dislodge them. As we have seen, in its present form the CEP does
not erode sovereignty per se, but rather challenges traditional patterns of
rule. In part, this challenge involves cultivating nonstate actors as sources
of knowledge generation, value transmission, and policy change. At the
same time it involves a seemingly contradictory effort to use state structures as decisive sources of leverage. By sanctioning a leading role for

states in framing the agenda and conducting collaborative work, the CEP in fact serves as a bulwark for state authority and functional capacity, at least in the aggregate.

Is this a desirable outcome from the standpoint of ecological security? The answer to this question is somewhat ambiguous. Many analysts have observed that the state itself—grounded as it is in the liberal international political economy—is inherently geared to extraction and competition and therefore is antithetical to the achievement of sustainable practices.[83] And yet it should be recognized that enforcement of environmental agreements often requires the exertion of institutionalized power, something that is absent in decentralized transnational movements or local communities.[84] Although the specific manifestations of state control may be relatively benign or coercive,[85] the expectation of enforcement does much to promote compliance. In sum, post-Westphalian organization may be conducive to social and normative developments consistent with "deep ecology" yet may also suffer shortcomings in terms of effective implementation. The CEP may prove to be a viable compromise between two extremes.

In addition, currents of modernization together with ecosystemic pressures progressively strain the state's ability to provide public goods. Inasmuch as centralized states are increasingly challenged as authoritative agents of political mediation and change, post-Westphalian forms of organization hold promise in the Caspian basin and elsewhere. Rather than eclipsing the state, these national and transnational policy networks provide a flexible, and—at least on an interim basis—complementary source of regulation and decision-making. With its simultaneous reliance on and transcendence of state structures, the evolving Caspian environmental regime represents an adaptive formulation of governance that fits within the growing interstices of state power.

Notes

1. Ken Conca, "Environmental Cooperation and International Peace," in Paul F. Diehl and Nils Petter Gleditsch, eds., *Environmental Conflict* (Boulder: Westview, 2000).

2. For a general overview, see A. Mekhtiev and A. Gul, "Ecological Problems of the Caspian Sea and Perspectives on Possible Solutions," in M. Glantz and I. Zonn, eds., *Scientific, Environmental, and Political Issues in the Circum-Caspian Region* (Dordrecht: Kluwer, 1997), 79–96; also see Rory Cox and Doug Norlen, *The Great Ecological Game: Will Caspian Sea Oil Development Lead to Environmental Disaster?* (Oakland, Calif.: Pacific Environment and Resources Center, January 1999).

3. The decline of the sturgeon population is also partly attributable to the destruction of natural spawning grounds due to river damming. Dmitry Pavlov, "Access to Spawn-

ing Grounds and Natural Reproduction in Caspian Acipenseridae," in H. Dumont, S. Wilson, and B. Wazniewicz, eds., *Caspian Environment Program: Proceedings from the First Bio-Network Workshop, Bordeaux, November 1997* (Washington, D.C.: World Bank, 1998), 1–9.

4. On fisheries problems see Igor Zonn, *Kaspiiskii Memorandum* (Moscow: Russian Academy of Natural Sciences, 1997), 137–68; and Igor Mitrofanov, "The Biodiversity of (Non-Sturgeon) Fish in the Caspian Sea," in Dumont, Wilson, and Wazniewicz, eds., *Caspian Environment Program,* 119–29. The economic importance of the industry is underscored, for example, in G. Voitolovskii, "Kaspiiskii uzel," *Rybnoe khoziaistvo,* no. 1 (January 1999): 8–12.

5. Philip Pryde, "Russia: An Overview of the Federation," in Philip Pryde, ed., *Environmental Resources and Constraints in the Former Soviet Republics* (Boulder: Westview, 1995), 25–39; and D. J. Peterson, "Russia's Environment and Natural Resources in Light of Economic Regionalization," *Post-Soviet Geography* 36, no. 5 (May 1995): 291–309.

6. Interviews at the Caspian Environmental Program, Program Coordinating Unit, and Thematic Center for Pollution Control, Baku, Azerbaijan, May 1999. See also "Half Billion Tons of Waste Discarded into Caspian Every Year," *Azernews/Azerkhabar,* February 17–23, 1999, in Azerbaijan News Distribution List, available through habarlar-1@usc.edu.

7. Faig Askerov and L. Rogers, "Oil Consortium Environmental Activity in the Caspian Sea," in Dumont, Wilson, and Wazniewicz, eds., *Caspian Environment Program,* 134–41; and Gilbert Rowe, "Azerbaijan, Oil, and Sustainable Development in the Caspian Sea," *Quarterdeck* 4, no. 3 (Winter 1996): 4–13.

8. A. Shcherbakova and W. Wallace, "The Environmental Legacy of Soviet Peaceful Nuclear Explosions," *CIS Environmental Watch,* no. 4 (Summer 1993): 33–56, at 43; see also *Aziyah-Ezh* (Almaty), no. 9 (September 15, 1995): 7, in Foreign Broadcast Information Service (FBIS), *Daily Report: Central Eurasia,* September 25, 1995 (FBIS-SOV-95-185-S), 81.

9. For a general overview see Zonn, *Kaspiiskii Memorandum,* 227–62. A prominent argument that attempts to explain the recent stabilization in sea level, emphasizing the importance of the Kara Bogaz Gol basin and influx from external sources, is G. Golitsyn et al., "On the Present-Day Rise in the Caspian Sea Level," *Water Resources* 25, no. 2 (1998): 133–39. Yet the underlying explanation for cyclical change remains poorly understood.

10. Zonn, *Kaspiiskii Memorandum,* 13–26.

11. For example, see *Workshop: Environmental Impact of Main Oil Pipelines: Abstracts* (Baku: Azerbaijan University Press, 1997).

12. D. Ya. Ratkovich and L. V. Ivanova, "The Problem of Development of the Russian Coast of the Caspian Sea," *Water Resources* 25, no. 3 (1998): 316–22.

13. Martha B. Olcott, "The Caspian's False Promise," *Foreign Policy,* no. 111 (Summer 1998): 95–113; and Jahangir Amuzegar, "OPEC as Omen," *Foreign Affairs* 77, no. 6 (November–December 1998): 95–111.

14. See Paul B. Henze, "Boundaries and Ethnic Groups in Central Asia and the Caucasus: Cause of Conflict and Change?" *Caspian Crossroads* 3, no.1 (Summer 1997), at ⟨ourworld.compuserve.com/homepages/usazerb/casp.htm⟩.

15. North Atlantic Treaty Organization (NATO), Committee on the Challenges of the Modern Society, "Caspian Risk Assessment," pilot study, March 1998 (photocopy).

On the underlying analytical connections between environmental degradation and overt conflict, see Thomas F. Homer-Dixon, "Environmental Scarcities and Violent Conflict: Evidence from Cases," *International Security* 19, no. 1 (Summer 1994): 5–40; and Günther Bächler, "Why Environmental Transformation Causes Violence: A Synthesis," *Environmental Change and Security Project Report* 4 (Spring 1998): 24–44. For an argument criticizing the putative causal linkage between environmental decay and conflict, see Marc A. Levy, "Is the Environment a National Security Issue?" *International Security* 20, no. 2 (Fall 1995): 35–62.

16. For a review of the legal issues, see especially Cesare Romano, "The Caspian and International Law: Like Oil and Water?" in William Ascher and Natalia Mirovitskaia, eds., *The Caspian Sea: A Quest for Environmental Security* (Dordrecht: Kluwer, 2000), 145–61; also see Bernard H. Oxman, "Caspian Sea or Lake: What Difference Does It Make?" *Caspian Crossroads* 2 (Winter 1996): 1–12.

17. For background on the scramble for oil and pipeline access, see Rosemarie Forsythe, *The Politics of Oil in the Caucasus and Central Asia,* Adelphi Paper no. 300 (London: International Institute of Strategic Studies, 1996); Suha Bolukbasi, "The Controversy over the Caspian Sea Mineral Resources: Conflicting Perceptions, Clashing Interests," *Europe-Asia Studies* 50, no. 3 (May 1998): 397–414; Amy Myers Jaffe and Robert Manning, "The Myth of the Caspian's Great Game: The Real Geopolitics of Energy," *Survival* 40, no. 4 (Winter 1998–99): 112–29; and Shirin Akiner and Anne Aldis, *The Caspian: Politics, Energy, Security* (New York: St. Martin's, 2000).

18. Elkhan Nuriyev, "Regional Conflicts and the New Geopolitics of NATO Expansion: The Cases of the Caucasus," paper presented at the annual convention of the American Association for the Advancement of Slavic Studies, Seattle, November 1997; and V. A. Teperman, "Strany Zakavkazia mezhdu Rossiei i iuzhnymi sosediami," in A. M. Khazanov et al, eds., *Rossiia, Blizhnee i Dalnee Zarubezhe Azii* (Moscow: Russian Academy of Sciences, 1997), 29–37.

19. Stephen Kinzer, "A New Big-Power Race Starts on a Sea of Crude," *New York Times,* January 24, 1999, sec. 4, p. 4. I discuss these attitudes in greater depth in Douglas Blum, "Sustainable Development and the New Oil Boom: Cooperative and Competitive Outcomes in the Caspian Sea," Program on New Approaches to Russian Security working paper no. 4 (Washington, D.C.: Center for Strategic and International Studies, May 1998).

20. James Richter, "Russian Foreign Policy and the Politics of National Identity," in Celeste Wallander, ed., *The Sources of Russian Foreign Policy after the Cold War* (Boulder: Westview, 1996), 69–94; and Vera Tolz, "Forging the Nation: National Identity and Nation Building in Post-Communist Russia," *Europe-Asia Studies* 50, no. 6 (1998): 993–1022. See also Sergei Kortunov, *Russia's National Identity in a New Era* (Cambridge. Mass.: Harvard University, J.F.K. School of Government, Strengthening Democratic Institutions Project, September 1998).

21. Robert O. Freedman, "Russia and Iran: A Tactical Alliance," *SAIS Review* 17, no. 2 (1997): 93–109; and Jahangir Amuzegar, "Iran under New Management," *SAIS Review* 18, no. 1 (Winter/Spring 1998): 73–92.

22. On regional extractive interests and political competition as a bar to environmental cooperation, see Mary Matthews, "International Finance Organizations and Incentives for Regional Environmental Cooperation in the Black and Caspian Sea Regions," paper presented at the meeting of the International Association for the Study of Common Property, Bloomington, Indiana, June 2000. On the influence of global

geopolitics, see Douglas Blum, "Environmental Change and U.S. Policy in the Caspian Sea," in Paul Harris, ed., *Environmental Change and U.S. Foreign Policy* (Oxford: Oxford University Press, 2001); and Ian Bremmer, "Oil Politics: America and the Riches of the Caspian Basin," *World Policy Journal* 15, no. 1 (Spring 1998): 27–35.

23. A secondary factor may well have been a renewed Iranian interest in pragmatic foreign relations with neighboring states, especially in light of the ongoing U.S. policy of "dual containment." See Houman Sadri, *Revolutionary States, Leaders, and Foreign Relations* (Westport: Praeger, 1997), 87–114.

24. A chronology of relevant meetings through May 1998 is included as an appendix to the "Concept Paper" cited in note 28.

25. For further discussion on this point see Douglas Blum, "Domestic Politics and Russia's Caspian Policy," *Post-Soviet Affairs* 14, no. 2 (April–June 1998): 137–64; for the broader context of this Russian approach, see Dina Zisserman, "The Politicization of the Environmental Issue within the Russian Nationalistic Movement," *Nationalities Papers* 26, no. 4 (December 1998): 677–86. For a comparative perspective see Miranda Schreurs, "Domestic Institutions and International Environmental Agendas in Japan and Germany," in Miranda Schreurs and Elizabeth Economy, eds., *The Internationalization of Environmental Protection* (Cambridge, U.K.: Cambridge University Press, 1997), 134–61.

26. For a discussion of this point see Miriam Lowi, *Water and Power* (Cambridge, U.K.: Cambridge University Press, 1993).

27. Telephone interviews with Phillip Tortell (UNDP), December 1998; Herb Behrstock (UNDP), December 1998; Andy Hudson (Global Environment Facility [GEF]), January 1999; Lawrence Mee (GEF), November 1998; and Amy Evans (World Bank), January 1999; and interview with Frits Schlingemann (UNEP), Moscow, May 1999.

28. For a general overview see CEP, "Concept Paper," May 3, 1998. Throughout this paper I refer to this as the "Concept Paper" for the sake of simplicity. Interested readers may find additional information in the GEF project brief "Addressing Transboundary Environmental Issues in the Caspian Environmental Programme": ⟨www.gefweb.org/wprogram/Oct98/WP_Sum.htm⟩. An informative website for the CEP can be found at ⟨www.caspianenvironment.org/⟩.

29. The literature on regimes is vast, and the debate over their effectiveness is ongoing. On regime effectiveness in reshaping preferences, see especially Robert O. Keohane and Lisa Martin, "The Promise of Institutional Theory," *International Security* 20, no. 1 (Summer 1995): 39–52; and for an overview of regime theory, see Andreas Hasenclever, Peter Mayer, and Volker Rittberger, *Theories of International Regimes* (Cambridge, U.K.: Cambridge University Press, 1997). On relative gains concerns and traditional state interests as constraints on regime performance, see Joseph Grieco, *Cooperation among Nations: Europe, America, and Non-Tariff Barriers to Trade* (Ithaca: Cornell University Press, 1990).

30. "Concept Paper," articles 60b and 60d.

31. Interview with David Aubrey, Woods Hole, Mass., October 1998.

32. "Concept Paper," article 60a.

33. On regime design and effectiveness, see Kenneth Oye, ed., *Cooperation under Anarchy* (Princeton: Princeton University Press, 1986); on the importance of transparency and problems in achieving it see Ronald Mitchell, "Sources of Transparency: Information Systems in International Regimes," *International Studies Quarterly* 42, no. 1 (March 1998): 109–30.

34. On the importance of diffuse reciprocity for broadening cooperation, see Robert Keohane, "Reciprocity in International Relations," *International Organization* 40, no. 1 (Winter 1986): 1–27.

35. Both objectives are emphasized by senior officials in private discussion. On capacity building, see "Concept Paper," articles 30–32.

36. "International Ecological Program to Begin in Spring on Caspian," *RIA Novosti,* January 28, 1998; "Caspian States Say Oil Projects Must Be Environmentally Sensitive," *FSU Oil & Gas Digest,* December 1998; and E. Kamiloglu, "Caspian Soap Opera Continues in Predictable Fashion: Iran and Russia Express Opposition to Transcaspian Oil and Gas Pipelines," *Zerkalo* (Baku), July 25, 1998, 6.

37. Telephone interview with Mark Berman (UNEP), January 1999; interview with Schlingemann. See also *Draft Framework Convention for the Protection of the Marine Environment of the Caspian Sea* (Nairobi: UNEP, December 1998).

38. Matthews, "International Financial Organizations."

39. To some extent this may be a result of third-party intervention by the international organizations as well as of a low-key, depoliticized approach to negotiations themselves. See P. Terrence Hopmann, *The Negotiation Process and the Resolution of International Conflicts* (Columbia: University of South Carolina Press, 1998).

40. This problem is widely recognized within the CEP network. See also "Concept Paper," article 71.

41. Interviews with Aubrey, Evans, and Hudson.

42. Conca, "Environmental Cooperation and International Peace."

43. See Karen Dokken, "Environmental Conflict and International Integration," in Nils Petter Gleditsch, ed., *Conflict and the Environment* (Dordrecht: Kluwer, 1997), 519–34.

44. Larry Swatuk, "The Environment, Sustainable Development, and Prospects for Southern African Regional Cooperation," in Larry A. Swatuk and David R. Black, eds., *Bridging the Rift: The New South Africa in Africa* (Boulder: Westview, 1997), 127–52; see also Peter Haas, *Saving the Mediterranean: The Politics of International Environmental Cooperation* (New York: Columbia University Press, 1990).

45. Such standards have existed since the fall of the Soviet Union but are not observed. They are currently being refashioned within the CEP context. On the design requirements of fisheries regimes, see Peter Sand, *The Effectiveness of International Environmental Agreements: A Survey of Existing Legal Instruments* (Cambridge, U.K.: Cambridge University Press, 1992). See also Natalia Mirovitskaya and J. Christopher Haney, "Fisheries Exploitation as a Threat to Environmental Security: The North Pacific Ocean," *Marine Policy* 16, no. 5 (July 1992): 243–58. On quotas see Martijn Wilder, "Quota Systems in International Wildlife and Fisheries Regimes," *Journal of Environment and Development* 4, no. 2 (Summer 1995): 55–104.

46. On development of these guidelines see "Concept Paper," articles 41–51.

47. For elaboration see Douglas Blum, "National, Sub-National, and International Politics: Environmental Regime-Building in the Caspian Sea," in Ascher and Mirovitskaia, eds., *The Caspian Sea,* 313–26.

48. Pipelines could be regulated for location, thickness, depth, monitoring technology, and bed structure.

49. A good background analysis of the regional politics associated with the Russian naval fleet, as well as a proposal for sharing naval capabilities for cooperative uses, is

found in Aleksei Vasilev, "O sviashchennoi korove: Po povody roli Kaspiiskoi flotilii," *Volga,* November 24, 1998.

50. Robert Cutler, "Cooperative Energy Security in the Caspian: A New Paradigm for Sustainable Development?" *Global Governance* 5, no. 2 (April–June 1999): 251–71.

51. For a similar argument on the need to balance ecology and navigation see D. M. Bodansky, "Protecting the Marine Environment from Vessel-Source Pollution: UNCLOS III and Beyond," *Ecology Law Quarterly* 18, no. 4 (1991): 764–77.

52. The ability of expert consensus to shift policy preferences by demonstrating causal links between actions and outcomes has been well documented. See the articles on epistemic communities in *International Organization* 46, no. 1 (Winter 1992); also see Thomas Risse-Kappen, "Ideas Do Not Float Freely: Transnational Coalitions, Domestic Structures, and the End of the Cold War," *International Organization* 48, no. 2 (Spring 1994): 185–214.

53. As Karen Litfin demonstrates, the manipulation of scientific knowledge is a discursive political process undertaken by interested and ideologically privileged actors. Karen Litfin, *Ozone Discourses: Science and Politics in Global Environmental Cooperation* (New York: Columbia University Press, 1994).

54. Marc A. Levy, "European Acid Rain: The Power of Tote-Board Diplomacy," in Peter M. Haas, Robert O. Keohane, and Marc A. Levy, eds., *Institutions for the Earth: Sources of Effective International Environmental Protection* (Cambridge, Mass.: MIT Press, 1993). See also Oran Young, "Negotiating an International Climate Regime: The Institutional Bargaining for Environmental Governance," in Nazli Choucri, ed., *Global Accord* (Cambridge, Mass.: MIT Press, 1993): 431–52; and O. A. Sankoh, "Making Environmental Impact Assessment Convincible to Developing Countries," *Journal of Environmental Management* 47, no. 2 (June 1996): 185–90.

55. According to the "Concept Paper," article 60d, "the linkages between Caspian Regional Expert Centers will form the backbone of the CEP."

56. The first meeting was held in Bordeaux in November 1997. For details see Dumont, Wilson, and Wazniewicz, eds., *Caspian Environment Program.*

57. Peter Willetts, *The Conscience of the World: The Influence of Nongovernmental Organizations in the UN System* (Washington, D.C.: Brookings Institution Press, 1996). Litfin emphasizes the centrality of popular pressure channeled by NGOs as the "driving force" behind all "apparently state-centric activities" in international environmental cooperation. Karen Litfin, "Sovereignty in World Ecopolitics," *Mershon International Studies Review* 41, no. 2 (November 1997): 167–204, quote at 175.

58. "Concept Paper," article 31.

59. "Addressing Transboundary Environmental Issues," sec. 3, H, 46. See also "Concept Paper," article 35. One member of the PCU staff was originally intended to act as liaison with NGOs throughout the region. Interview with Stuart Gunn (Baku PCU), Baku, November 1998.

60. Telephone interviews with Evans; Tortell; Behrstock; David Vousden (UNDP), January 1999; and Marc Wilke (TACIS), January 1999.

61. On how hegemonic world political and conceptual structures tend to limit change and to cause environmental pressures to be addressed in such a way as to reproduce those existing structures, see Ken Conca, "Environmental Change and the Deep Structure of World Politics," in Ronnie D. Lipschutz and Ken Conca, eds., *The State and Social Power in Global Environmental Politics* (New York: Columbia University Press, 1993), 306–26.

62. Interviews with Bakhtiar Muradov (Baku PCU), Baku, May 1999; Kamran Abdullaev ("Ecoil Ruzgar"), Baku, May 1999; and Arkadiusz Labon (Fisheries Thematic Center), Astrakhan, May 1999. See "Experts View Caspian Oil Extraction Aspects," *Delovaya nedelya* (Almaty), September 4, 1998, trans. in Azerbaijan News Distribution List, September 26, 1998. NGO representatives took part in the Steering Committee meeting held in Baku, June 1999.

63. It is possible to identify cases of such public activism. See, for example, "Azeri Ecologists against Oil Companies Discharging Sludges into Caspian Sea," Turan news agency (Baku), in Russian, March 6, 2001, reprinted in *BBC Monitoring,* March 8, 2001; and "Kazakh Greens Say Drilling in Caspian Sea Damages Environment," Interfax-Kazakhstan news agency (Almaty), in Russian, February 1, 2001, reprinted in *BBC Monitoring,* February 22, 2001. A highly (and, in my view, excessively) pessimistic view of Caspian environmental NGOs is found in Eric Sievers, "How NGOs Abandoned Governance in the Caspian Region," in Ascher and Mirovitskaia, eds., *The Caspian Sea,* 219–33.

64. ISAR has launched a major campaign to facilitate such contacts and increase direct public involvement, and toward this end it has held several regional NGO conferences. See the articles in the special issue of ISAR's journal on "Reaching Out to Protect the Caspian": *Give & Take* 3, no. 4 (Winter 2001). I am indebted to Kate Watters of ISAR for many helpful discussions of these issues.

65. For a positive assessment see Ian Bowles, "The Global Environment Facility: New Progress on Development Bank Governance," *Environment* 38, no. 3 (April 1996): 38–41; and, for a more critical assessment, Rodger Payne, "The Limits and Promise of Environmental Conflict Prevention: The Case of the GEF," *Journal of Peace Research* 35, no. 3 (May 1998): 363–81. Also, telephone interview with Mee (the former director of the Black Sea Environmental Program, UNDP). Like the CEP, the Soros Foundation has launched an educational initiative. "Soros Foundation Organizes International Ecological Movement to Protect Caspian," *Assa-Irada* (Azerbaijan International Independent News Agency), August 28, 1998.

66. "Addressing Transboundary Environmental Issues," sec. 3, H, 46, and sec. 5, 1, A, 62–63; see also "Concept Paper," article 31. For a concrete example see "Seminars within Caspian Ecological Program Due in February," *Azernews-Azerkhabar,* January 13–18, 1999.

67. On the potential for refashioning national collective identities on the basis of constitutive norms, see Alexander Wendt, "Collective Identity Formation and the International State," *American Political Science Review* 88, no. 2 (June 1994): 384–98.

68. For a thoughtful analysis of this process see Ronnie Lipschutz with Judith Mayer, *Global Civil Society and Global Environmental Governance* (Albany: State University of New York Press, 1996).

69. "Segodnia v tegerane nachnet rabotu ezhegodnaia konferentsiia po problemam Kaspiiskogo moria," *Novosti,* June 22, 1999, available at ⟨www.rian.ru/acsna/r_acsna. htm⟩. See also the remarks by former Russian minister for CIS relations, Boris Pastukhov, in "Strasti po chernomu zolotu," *Moskovskii komsomolets,* November 12, 1998; and "Interview with Deputy Minister of Foreign Affairs, Special Representative of the President of Russia on questions of regulation of the Caspian Sea, V. I. Kaliuzhnyi," informational bulletin (in Russian), Russian Ministry of Foreign Affairs, October 2, 2001, available at ⟨www.mid.ru/⟩.

70. Jeffrey Checkel, "Norms, Institutions, and National Identity in Contemporary Europe," *International Studies Quarterly* 43, no. 1 (March 1999): 83–114.

71. See Thomas Biersteker and Cynthia Weber, eds., *State Sovereignty as Social Construct* (Cambridge, U.K.: Cambridge University Press, 1996); and Conca, "Environmental Cooperation and International Peace."

72. Paul Wapner, "Politics beyond the State: Environmental Activism and World Civic Politics," *World Politics* 47, no. 3 (April 1995): 311–40; and Lipschutz, *Global Civil Society.*

73. For a brilliant discussion of the historical conditions that have spawned successive principles of governance, as well as the international political implications of such changes, see Rodney Bruce Hall, *National Collective Identity: Social Constructs and International Systems* (New York: Columbia University Press, 1999).

74. On the decline of the state's traditional social welfare role under conditions of globalization, see John Ruggie, "At Home Abroad, Abroad at Home: International Liberalization and Domestic Stability in the New World Economy," *Millennium: Journal of International Studies* 24, no. 3 (Winter 1995): 507–26. On the state's inability to act entirely apart from the sanctioning body or to claim independent authorship for the benefits conveyed via international cooperation, see Ken Conca, "Rethinking the Ecology-Sovereignty Debate," *Millennium: Journal of International Studies* 23, no. 3 (January 1994): 701–11.

75. Litfin, "Sovereignty in World Ecopolitics."

76. NGOs may thus provide an independent source of accountability or may simply enact delegated functions. On NGOs' strengthening of state capacity, see Kal Raustiala, "States, NGOs, and International Environmental Institutions," *International Studies Quarterly* 41, no. 4 (December 1997): 719–40. See also Ann Hawkins, "Contested Ground: International Environmentalism and Global Climate Change," in Lipschutz and Conca, eds., *The State and Social Power,* 221–45, esp. 237–40. Thanks to Ken Conca for discussion of this point.

77. Interviews with Aubrey, Evans, and Schlingemann.

78. Additionally, national focal points "liaise with the regional PCU and their country's representative to the Regional Steering Committee." "Concept Paper," article 60c.

79. Anne-Marie Slaughter, "The Real New World Order," *Foreign Affairs* 76, no. 5 (September/October 1997): 183–97; see also Jan Kooiman, ed., *Modern Governance: New Government-Society Interactions* (London: Sage, 1993).

80. Such fears were evident in private discussions with CEP officials from Turkmenistan, Kazakhstan, and Azerbaijan in 1998 and 1999.

81. Such fears were displayed by the Iranian delegation at a meeting of the Thematic Center for Institutional, Legal, Regulatory, and Economic Instruments of Environmental Management, held in Moscow, May 1999.

82. For these and other cogent concerns about the CEP's viability, including the declining level of science in the littoral states, see Eric Sievers, "Caspian Environment Programme: Prospects for Regime Formation and Effectiveness," in Ascher and Mirovitskaia, eds., *The Caspian Sea,* 327–43.

83. See, for example, Richard Falk, *On Humane Governance: Towards a New Global Politics* (Cambridge, U.K.: Polity, 1995); and Wolfgang Sachs, "Global Ecology and the Shadow of Development," in Wolfgang Sachs, ed., *Global Ecology* (London: Zed, 1993), 3–21.

84. M. J. Peterson, "International Organizations and the Implementation of Environmental Regimes," in Oran Young, ed., *Global Governance: Drawing Insights from the Environmental Experience* (Cambridge, Mass.: MIT Press, 1997), 115–51.

85. For an argument against the international environmental community's inadvertent support for state coercion and its negative moral (and possibly practical) consequences, see Nancy Peluso, *Rich Forests, Poor People: Resource Control and Resistance in Java* (Berkeley: University of California Press, 1994).

7

Water Cooperation in the U.S.-Mexico Border Region

Pamela M. Doughman

The absence of violence is only the beginning of the work for peace.

—Nobel laureate Betty Williams[1]

People should work out their own priorities, and the self-reliant ones will always have the courage to do so in dialogue with others.

—Johan Galtung[2]

One theme of this volume is that international environmental policies should be assessed not only in terms of their potential ecological impact but also in terms of their impact on peaceful cooperation among nations and across societies. Many of the chapters in this volume involve world regions facing chronic insecurities and political instability, in which peace is often tenuous and large-scale violent conflict is all too easy to imagine. But we should also ask whether an "environmental peacemaking" framework can shed light on relations in settings where violence and insecurity, if no less real, are somewhat more diffuse. This chapter examines whether this perspective can be applied usefully to the United States and Mexico— specifically, to the growing institutionalization of environmental protection initiatives in the border region.

The U.S.-Mexico border region is not a war zone, but neither is it a zone of peace. The border is an area characterized by environmental degradation, economic and political inequality, and conflicting norms and priorities. A knowledge struggle is underway regarding the nature and importance of environmental quality along the border. As much of the border area is arid land, environmental tensions tend to center on water quality and supply, although air quality and solid-waste management are also worrisome.

190

This chapter argues that water cooperation efforts in the U.S.-Mexico border region have the potential to improve U.S-Mexico relations in many respects. These efforts contain processes that reduce uncertainty, promote more diffuse forms of reciprocity, and encourage long-term planning by governing bodies. They also institutionalize processes that create new forms of interdependence, promote norms of environmental protection, and strengthen trans-societal linkages. Although their performance to date has been uneven, these institutions could help to move the border from a zone of uneasy transition and human insecurity to a zone of peace.

The U.S.-Mexico Border: Perils and Promise

A visitor strolling through the foyer of the Museo de las Californias (Museum of the Californias) in Tijuana, Mexico, in June 2000 would have encountered a forest of large banners hanging from the ceiling. They had been generated by the second international competition for banners characterizing the U.S.-Mexico border. Violence permeated the art. One of the more subtle pieces centered on a dream-like photograph of an idyllic seashore, a fence, and a somber list of names—and no-names—of people who have died in the process of illegally crossing the U.S.-Mexico border through Tijuana.[3] Another banner depicted fourteen international conflicts ongoing at the time of the competition: Palestine, Colombia, Chechnya, Kashmir, Congo, Sierra Leone, Chiapas, Kosovo, Ecuador, Rwanda, Burundi, East Timor, Tijuana, and Cyprus.[4] The artist was appalled by the relative quiescence in the international community regarding these conflicts. He felt they were manifestations of a violent culture that emanates from the United States.[5] The collective image of these works conveys a need for improvement in the way that people encounter the U.S.-Mexico border. For many, the encounter is not as peaceful as it could be. And it is not only current conflicts that influence the peacefulness of the border. To many in Mexico, the memory of the Mexican-American War of 1848 continues to make U.S.-Mexico border relations problematic.

But this is not the whole story. The border is also a vibrant and beautiful place. The multidimensional Tijuana area, for example, is a vibrant landscape of abrupt transitions, complexity, and contrasts. Near the discussion of violence in the Museo de las Californias was a disarmingly inspiring mural of the landscape of the Tijuana valley: "Tierra Prometida" (Promised land), by Juan Angel Castillo. The mural conveys a sense of the vitality and possibility of Tijuana. The valley cannot be contained in one wall: the mural covers two walls at right angles to each other. The first impression one encounters of the border depends on the entrance one chooses, but eventually both messages are impossible to avoid.

The vitality and opportunity of the border region have attracted people from the United States, Mexico, China, Japan, and elsewhere to invest time, labor, and life in its landscape, economy, and society. At the same time, the border is plagued by problems associated with immigration, illegal substances, and hazardous waste. Efforts to address these problems—be they an electrified border fence, security checkpoints on major highways, or industrial manufacturing facilities—have often aggravated social, political, and ecological tensions at the border. For example, armed guards at a security checkpoint along a major highway outside of Tijuana have been assigned to pull over vehicles for inspection. The process can be quite disconcerting.

The environmental problems in the border region are formidable. Coastal waters nearest the border are contaminated by wastewater from the area's maquiladora facilities[6] and rapidly growing residential and migrant populations. The contamination often affects nearby U.S. beaches as well.[7] In addition, the border region suffers from toxic pollutants in the air, water, and soil. According to the environmental ministries of the two countries sharing the border,

> contamination of air, water, and soil by hazardous materials and waste, pesticides, nitrates, raw sewage, untreated wastewater, parasites, or bacteria are suspected to be key factors contributing to the presence of certain diseases in the populations residing along the border. These diseases include asthma and tuberculosis; elevated blood lead levels in children; multiple myeloma, a form of bone-marrow cancer; systemic lupus erythematosus, an autoimmune disorder; hepatitis A; infectious gastrointestinal diseases such as shigellosis and amebiosis; and pesticide poisonings.[8]

Scholars and residents point to the unequal distribution of economic, political, and semiotic power between the United States and Mexico as a driving factor behind the troubles of the border region. For example, residents of Tijuana complain that the social fabric of their city is being undermined by the influx of desperate people in search of employment in the United States.[9] Border residents do not believe that maquiladoras are reducing this instability; they point to the high employment turnover as evidence.[10] These views are consistent with the suggestion by Johan Galtung that one of the causes of insecurity in international relations is the existence of "exploitative relations between groups in general and societies in particular," an inequity referred to by Galtung and others as "structural violence."[11]

For Galtung, structural inequity is a problem because it interferes with human development, which he understands to mean "human self-realization and liberation."[12] To improve the process of advancing human development, Galtung suggests actor-oriented and structure-oriented activities that reduce violence, misery, repression, and alienation. To avoid being counterproductive, such activities should reflect residents' concerns.[13]

Such a dialogue is underway regarding the nature and importance of environmental quality along the border. On this topic, there are at least three conflicting views of the border environment, creating uncertainty and fluctuating levels of support for policies designed to improve border conditions. This uncertainty has the potential to generate misunderstandings, delay, and inaction between the United States and Mexico. It may cause opportunities for mutual gains to go unidentified.

Views on the nature and urgency of border-related environmental problems vary with geographic location; different groups do not share the same sense of urgency regarding border-related environmental problems. Thus, if real change is to occur, efforts to strengthen environmental cooperation in the border region must strive to understand and respect the range of priorities present among those who need to cooperate with each other.

Toward this end, a brief discussion of three of the dominant impressions of the border is presented in this chapter. To varying extents, all three of these views are reflected in two new institutions that address environmental infrastructure problems along the border: the North American Development Bank (NADBank) and the Border Environment Cooperation Commission (BECC). These institutions, products of the North American Free Trade Agreement (NAFTA), have the potential to reduce structural inequalities and to deepen participation, trust, and interdependence in the border region. Their performance to date has been uneven, however, with some activities doing more to improve interstate and trans-societal relations than others.

Support for strengthening environmental protection has been promised by Mexico's president, Vicente Fox: "We will make Mexico's environment, its water and forests, a national-security issue. We will turn around the concept of development to include the environment as a factor in economic and social decisions—not as a separate sector, but as an essential element in creating sustainable economic and social progress."[14] It remains to be seen whether this promise provides solely rhetorical capital or generates economic and political capital for improving environmental protection in Mexico. Either way, the ongoing political transition in Mexico is likely to provide the BECC and NADBank an opportunity to expand their contributions to peace-building through environmental cooperation in the border region.[15]

Views on the Border: Public Health, Wealth, and Inequality

The way people define a situation, the metaphors they use to explain its meaning, and the cultural biases of institutions granted authority over the matter delimit the range of acceptable social action in that environment.[16] Thus, present-day tensions along the U.S.-Mexico border cannot be fully comprehended without reference to historical events.

The Mexican-American War is particularly salient in this regard. For many, the war is not over, and the outcome is not entirely accepted. The land lost by Mexico in the war is the same area in which many Mexicans strive to find gainful employment today. There are organized conservative activists in the United States who also feel that the war is not over. But rather than cooperation, they construe the necessary response as one of separation, exclusion, and reduced mobility across the border. For example, in a feminist critique of conservative discourses regarding instability in the border region, Susan Mains highlights the terms used by the conservative community organization Voices of Citizens Together (VCT):

> The VCT video ["Invasion of the United States"] raises concerns about immigrant bodies generally in relation to illegality, mobility, and loss of sovereignty. As [VCT founder] Glenn Spencer states, "Being non-citizens apparently will not get in the way of those who are determined to retake California. They are now overwhelming the democratic process itself." The 'natural' cycles of the individual immigrant body, therefore, are depicted as threatening the US at a state and national scale. . . . [A] narration [at a later point in the film states,] "If southern California is to become an annex of Mexico, if it becomes Aztlan, who will be the Gatekeeper? Who is to control immigration into the rest of the US?" . . .

The loss of a dominant legal or national identity, therefore, is also represented as being intricately interwoven with the increased physical presence of immigrants, a potential loss of land and therefore, the nation.[17]

Such arguments were used to support Proposition 187 (the 1994 California referendum denying public services to undocumented aliens). They have also been used to support "Operation Gatekeeper," the name given to U.S. Border Patrol efforts to reduce the permeability of the U.S.-Mexico border.

In terms of efforts to strengthen collective identity and build trust through environmental cooperation, this legacy poses a problem. A resident academic and activist in the border region described the situation as follows:

The American side of the border was created by amputation from Mexican territory. The Rio Bravo is the sight of the amputation. It is difficult to work with a dismembered part of your anatomy, . . . but we must learn . . . because in the end, geographically we are destined to live as neighbors. Sometimes we are distant neighbors, others say that we are distinct, but it is certain that we are companions along the path of a single life in a shared geographic area with common problems: water, air, the transport of toxic substances and of hazardous waste.[18]

If the goal is to strengthen the peaceful qualities of the U.S-Mexico border, it is important to understand the definitional, metaphorical, and institutional meanings brought to bear on its governance. In public policy, the media, and private conversation, three of the views that dominate public perceptions of the border center on health, wealth, and inequality. The view that poverty, industry, population growth, and poor infrastructure along the border create public health problems for both the United States and Mexico seems to resonate most strongly among U.S. activists and policymakers. From the Federal District of Mexico, the view that the border is a relatively wealthy area seems to dominate public understanding. In contrast, people living in the border region on the Mexican side tend to see the border as a crossroads of inequity.

In the NAFTA debate, environmentalists opposed to the accord framed their concerns regarding expanded trade with Mexico in terms of public health. They argued that freer trade would encourage manufacturers to locate along the Mexican side of the border and, due to poor enforcement of environmental laws, would result in increased air and water pollution, threatening public health in both Mexican and U.S. territory.[19] The Nogales Wash, the New River, and the Tijuana River, among others, bring pollutants from Mexico to the United States.[20] At other points, pollution flows from the United States to Mexico.[21] The view articulated by U.S. activists and policymakers in the NAFTA debate tended to focus on cooperative efforts to generate resources to reduce pollution rather than divisive efforts to place blame on one side or the other. Conditions at the time spurred the American Medical Association to describe the border as "a virtual cesspool."[22] Any increase in pollution, regardless of its source, was construed as a problem for both countries.

Public-health constructions continue to dominate policy debates. For example, Brian Bilbray, a Democratic member of Congress representing a district including part of San Diego, used such arguments in his request for supplemental congressional funds to help clean up sewage problems in the Tijuana River:

> Bilbray displayed a photograph taken by satellite with special imaging equipment that showed the flow of contaminated water out [of] the Tijuana River into the ocean and San Diego Bay. "This is not a problem being created by the people in these neighborhoods," he said. "It is created by a foreign government." Bilbray asked his colleagues to approve the money [$500,000] "as a sign that the United States government will do what is necessary to protect its citizens." . . . Bilbray said the money would pay for a barrier in the Tijuana River to divert the raw sewage into the treatment plant built under an international agreement to stop a health problem that has persisted for decades.[23]

While the U.S. discourse construes the border as an impoverished region, Mexican governance activities reflect the view that the border is a wealthy area (see Table 7.1). Using this assessment as a justification, government assistance for municipal development tends to ignore the border region.[24] Due to their proximity to the United States, the Mexican border states tend to have the economic and political wherewithal to be relatively self-reliant. The United Nations Development Programme ranks Mexico 55th out of 174 in the "Human Development Index" of its *Human Development Report 2000,* placing it ahead of the Russian Federation but behind South Korea, Chile, or Poland. (A comparison of the United States and

Table 7.1

Percentage of Mexican Workers Receiving More than Five Times the Minimum Wage, by State, 1997

State	Percentage of workers	Border state?
Baja California	19.7	Border
Distrito Federal	17.5	
Nuevo Leon	16.1	Border
Baja California Sur	15.1	
Quintana Roo	14.0	
Chihuahua	13.3	Border
Sonora	12.5	Border
Queretaro	12.0	
Aguascalientes	11.4	
Tabasco	11.2	
Coahuila	11.1	Border
Tamaulipas	10.5	Border
Sinaloa	9.9	
Colima	9.6	
Mexico	9.5	
Nation as a whole	*9.5*	
Campeche	9.2	
Jalisco	9.2	
Durango	8.7	
Guanajuato	8.0	
Morelos	7.2	
Nayarit	6.9	
Yucatan	6.8	
San Luis Potosi	6.5	
Veracruz	6.2	
Zacatecas	6.2	
Michoacan	5.8	
Puebla	5.0	
Hidalgo	4.8	
Chiapas	4.5	
Tlaxcala	4.4	
Guerrero	4.3	
Oaxaca	2.7	

Source: Instituto Nacional de Estadistica, Geografia e Informatica (INEGI), "Estadisticas sociodemograficas: Estructura porcentual de la poblacion ocupada por grupos de ingreso segun entidad Federativa, 1997 (por ciento)" (Sociodemographic statistics: Percentage structure of the working population by income groups according to federal entity, 1997 [percent]), based on "Poblacion ocupada y su distribucion porcentual por entidad federativa segun grupos de ingreso por trabajo en salario minimo mensual," *ENADID: Encuesta Nacional de la Dinamica Demografica 1997: metodologia y tabulados* (National survey of demographic dynamics, 1997: methodology and tables), 1st ed. (Aguascalientes, Ags.: Instituto Nacional de Estadistica, Geografia e Informatica, 1999), 429.

Table 7.2

Human Development Statistics for the United States and Mexico

	United States	Mexico
HDI rank[1]	3	55
Life expectancy at birth, 1998 (years)	76.8	72.3
Adult literacy rate, 1998 (%)	99.0	90.8
Combined primary, secondary, and tertiary gross enrollment ratio, 1998 (%)	94	70
Gross domestic product per capita, 1998 (PPP, U.S.$)[1]	29,605	7,704

Source: United Nations Development Programme (UNDP), *Human Development Report 2000* (New York: Oxford University Press, 2000).
[1]For an explanation of the Human Development Index (HDI) and purchasing power parity (PPP), see the notes to Table 2.1 on page 26.

Mexico using statistics associated with the Human Development Index is shown in Table 7.2.)

Rather than health or wealth, residents on the Mexican side of the border tend to focus on inequality.[25] When asked to identify the greatest problem in the border region, one resident responded that before addressing environmental concerns, the gap between the rich and the poor must be narrowed.[26] Inequality is seen to function as a magnet for immigration through Mexico to the United States. In 2000, jobs that paid $5 for a day's work in Mexico paid $60 in the United States.[27] In this light, some border residents see the restrictions on Mexican migration into the United States for work as a cruel irony. According to this view, life may not necessarily be better for the would-be immigrants, but as long as they think it would be, it is cruel to deny them a chance to find out.[28]

This view was reflected in the Mexican television coverage of two people who died on June 7, 2000, trying to cross the Rio Grande near Matamoros, Tamaulipas, and Brownsville, Texas. U.S. and Mexican border agents were present at the drowning, but did not have the ability to save the victims.[29] A Televisa TV crew filming nearby captured the incident, which was broadcast throughout Mexico.[30] On June 11, a panel of four people was gathered by Tijuana's Channel 11 on their "Contra punto" [Counterpoint] program to comment on the situation. One called for the Mexican government to be better prepared to save the lives of people who appear to be drowning while trying to illegally cross the border. Another called for a temporary and seasonal worker program in the United States to reconcile the would-be immigrants' desperate search for employment with the U.S. interest in reducing illegal immigration. "We are sending young people to their deaths!" exclaimed one commentator. "Which of the [Mexican] presidential candidates have talked about this seriously?"[31]

Water issues also evoke complaints of inequity. One thorn of contention in the border region is the amount and salinity of Colorado River water reaching Mexico. A lawsuit was filed in 2000 by environmentalists in the United States as part of an effort to preserve endangered species in the Colorado River delta.[32] In a related matter, academics and environmentalists have raised calls of alarm regarding the harm to the environmental and agricultural systems in the Mexicali Valley of Baja California expected to result from the reduction in available groundwater once the concrete lining of the All-American Canal is completed.[33] Since 2000, a transboundary water controversy persists regarding the amount of water reaching Texan farmers from Mexican reservoirs feeding the Rio Grande; the farmers were not getting the amount they had been promised.[34]

Thus, geographical location tends to color the view of the border that appears most clearly on the horizon. All three of the views of the border region—as a public health problem, a self-reliant and comparatively wealthy area, and a crossroads of inequality—are embedded in the new development institutions created by NAFTA, the BECC and NADBank. The institutions reflect the assumption that border communities see wastewater treatment as a public health priority, that these communities have the economic resources to shoulder most of the infrastructure costs themselves, and that the contrasting social and economic conditions within and across U.S.-Mexico border communities necessitate a cooperative effort to understand and share experience and expertise.[35] If the tensions and contradictions of the U.S.-Mexico border pose an opportunity to strengthen trust, reciprocity, long-term planning, interdependence, shared norms, and trans-societal linkages, then these institutions may be well placed to help move the border from a zone of uneasy transition to a zone of peace.

The North American Development Bank and the Border Environment Cooperation Commission

In order to reduce trade barriers in North America, the governments of the United States, Canada, and Mexico began negotiating NAFTA in February 1991. Negotiations concluded on August 12, 1992. Support for expanded economic interdependence with Mexico was seriously weakened by concerns about its implications for environmental quality and labor practices. During the negotiations, environmental groups repeatedly communicated their concerns about the potentially negative environmental impact of a trade-liberalizing agreement with Mexico. Mexico had a very poor record of implementation and enforcement of its environmental laws. The worry was that companies might move to Mexico to avoid the costs associated with pollution-prevention requirements in the United States and Canada.[36] Organized labor was similarly concerned that lax labor standards and low wages in Mexico would lead to weakened labor standards and wage practices in the United States and Canada.[37]

In the midst of the NAFTA debate on these themes, Mexico won a case against a U.S. environmental law, the Marine Mammal Protection Act. A dispute-resolution panel of the General Agreement on Tariffs and Trade ruled against U.S. efforts to prevent the importation of tuna caught in ways harmful to dolphins.[38] The case enhanced political pressure for the inclusion of environmental protections in NAFTA. In order to broaden political support for expanded economic interdependence, the United States, Canada, and Mexico agreed to augment cooperation on both labor and environmental issues through side-agreements to NAFTA.

One of the outcomes of the NAFTA debate was the creation of two institutions designed to promote norms of environmental protection and strengthen trans-societal linkages. The new institutions were intended to expand the role that communities within 100 kilometers (km) of the U.S.-Mexico border play in determining and providing for their own environmental infrastructure needs.[39] The BECC was created to help border communities conceive and design environmental infrastructure projects, for which it provides certification of "technical feasibility and environmental integrity." The BECC has a ten-member board of directors, divided evenly between Mexico and the United States and including federal and local governmental officials as well as representatives of the private and public-nonprofit sectors.

The BECC's "sister institution," NADBank, is charged with financing the development of environmental infrastructure certified by the BECC. NADBank's board of directors consists of six cabinet-level officials from the two countries: the U.S. secretary of the Treasury, secretary of state, and administrator of the Environmental Protection Agency, and the Mexican secretary of finance and public credit, secretary of commerce and industrial development, and secretary of social development. NADBank describes its services to border communities as those of adviser and financial strategist, investment banker, and lender. To ensure financial self-sufficiency, NAD-Bank is required to lend money at above-market interest rates.

As of July 2000, the BECC had certified 40 projects, 29 of which had been financed by NADBank.[40] Projects range from the $80.4 million comprehensive sanitation project in Ciudad Acuna in the state of Coahuila to the $155,000 on-site wastewater treatment system self-help loan project for the *colonias* of El Paso County, Texas. The total value of the 29 projects financed by NADBank is $831 million. NADBank has contributed more than $264 million toward these loans ($253 million from the NADBank-administered Environmental Protection Agency Border Environment Infrastructure Funds and $11.2 million in NADBank loans).

Controversies: The Pace of Lending and the Scope of NADBank's Mandate

Lending to date constitutes only a small portion of NADBank's $2 billion lending capacity.[41] In August 2001, Jorge Garces, deputy managing director of NADBank, was critical of the bank's lending performance: "We are the first to admit that our lending record is very, very poor. Yes, in a sense, we have failed miserably but that's because of the interest rate situation. It has been that way since we were set up."[42]

The money sitting unused in NADBank has become the target of rival organizations with conflicting views regarding the nature and importance of water management problems in the border region. In Mexico, Hacienda (the Treasury) and SECOFI (the trade and economic agency) have stated that NADBank's success is a function of the amount of money it loans. The Border Trade Alliance, the President's Inter-Agency Task Force on the Economic Development of the Southwest Border, and local governments are ready to find useful projects for the funds should they be released from NADBank control.[43] Others argue that funds should be used to promote economic development in the home states of people moving to or through the border area looking for work. For example, Cyrus Reed of the Texas Center for Policy Studies (TCPS), an NGO advocating improved environmental and community health in the border region, argues that "NADBank should be helping those regions of Mexico where economic opportunity is so scarce that people are forced to migrate in search of work. We may have less illegal immigration if they did."[44]

Efforts to resolve this conflict have increased information regarding stakeholders' preferences, intentions, and the root causes of NADBank's troubles. The discussion has helped develop shared norms and priorities among interested social actors. It has also helped identify new opportunities for mutual gain and cooperation. In a resolution approved by the NADBank board of directors on July 11, 2000, the bank argued that it is undesirably constrained from making a greater contribution to human health and environmental concerns in the U.S.-Mexico border region by its limited mandate:

> The BECC and NADB have successfully assisted and funded the development and construction of 29 water, wastewater and solid waste projects, but the loan components of their financial packages have been minimal, due to legal barriers in the United States, the availability of subsidized lending in the United States and, in particular, community and market conditions in both the United States and Mexico.[45]

In accordance with this resolution, in July 2000, NADBank released a draft document titled "Utilizing the Lending Capacity of the NADB" and opened a forty-five-day public comment period on whether and how NADBank should expand the types of infrastructure it supports and the geographic region it serves. At issue was whether NADBank activities should be expanded to 300 km on either side of the border; whether the types of projects NADBank can fund should be expanded beyond the water, wastewater, and solid waste sectors; and whether there are alternative financial options it should be allowed to offer to make its loans more affordable to border communities.[46]

Early in 2001, NADBank's mandate was expanded to include projects in the following areas: water conservation, water and wastewater residential connections, recycling, waste reduction, and industrial and solid waste projects that affect actual or potential water and soil pollution. In September 2001, Fox and U.S. President George W. Bush discussed reform of NADBank and the BECC. At the meeting, the presidents agreed that "immediate measures were needed to strengthen the performance" of NADBank and the BECC "to identify and fund environmental infrastructure projects on the border." Toward this end, they stated that they would hear the recommendations of a bi-national working group on this topic by October 31, 2001.[47] The issue of expanding NADBank's mandate is a vehicle through which both state and nonstate actors can consider whether to commit to a more diffuse form of reciprocity in the area of environmental cooperation at the U.S.-Mexico border. The arguments made during the public debate over the expansion of NADBank's mandate highlight conflicting interpretations of priorities at the border. They also illustrate the process through which identification of opportunities for further collaboration have emerged.

Admonishing an audience of San Diego's elite that their "dreams need to be more aggressive," Victor Miramontes, a leading NADBank figure at the time, suggested that the 150-mile distance from the Imperial and Mexicali valleys to San Diego should be imagined as a more contiguous unit, enhanced by a high-speed commuter rail financed by a NADBank with the authority to support border transportation projects: "In Miramontes' vision of the future, factories would multiply in Mexicali, where there's an abundance of water and electricity. Middle-income families, priced out of San Diego and Tijuana markets, would find enough affordable housing in the Imperial and Mexicali valleys to absorb development through the next century."[48]

NADBank's proposal to expand into new areas was met with a mixed reception. In July of 2000 the TCPS sponsored a workshop with representatives from the BECC, other NGOs, and academia in attendance. One of the conclusions reached at the workshop was that prior to expanding to new areas, the BECC and NADBank should conduct an evaluation to learn the effectiveness of the forty projects that have received BECC certification: Are they meeting the needs of the border residents they are supposed to benefit?[49]

In a white paper on NADBank's proposal, the TCPS supported the idea that NADBank should search for ways to make its loans more affordable to border communities. However, the TCPS opposed expansion of NADBank activities into new sectors or an expanded geographic area. The TCPS also focused on two issues absent from the NADBank proposal: the importance of BECC certification of all NADBank projects and the need for consideration of the cumulative impact of NADBank activities on the border region.

In stating its position against expansion into transportation and other new sectors (e.g., mortgages, water transfers, paving of roads), the TCPS put forward several arguments. First, neither NADBank nor the BECC has the resources or expertise to evaluate projects in such sectors. Second, it is likely that a reframing of the NADBank and BECC charters would require approval by the executive and legislative branches of both countries. Third, NADBank and the BECC have not exhausted the possibilities contained in the existing mandate covering water, wastewater, and solid waste and related matters.[50]

The TCPS also opposed the NADBank proposal to expand its jurisdiction to projects within 300 km of the border. Expanding the "borders" of the border region from 100 km to 300 km would bring wealthy metropolitan areas in competition with the poor communities located closer to the border, the group argued. In the proposed 300 km region, Los Angeles, Phoenix, San Antonio, and Monterrey would compete with Tecate, Heber, Brawley, and Aguas Prietas for funding.[51]

Expansion of NADBank's geographic area of operation has been linked to an effort to slow emigration of people from Mexico by creating jobs at home. In the September 2001 meeting between Fox and Bush, NADBank was separated from the issues of economic development in Mexico, a temporary worker program, and the "human dignity of all migrants" from Mexico to the United States. Rather than assigning NADBank the task of grappling with these issues, a new "public-private alliance to spur private sector growth throughout Mexico" was launched. The alliance was named the "Partnership for Prosperity." A deadline of March 1, 2002, was set for presentation of a "concrete plan of action" to promote the aims of the initiative.[52]

Controversies: The Role and Focus of the BECC

In addition to the geographic jurisdiction of NADBank, the TCPS was concerned by the lack of specificity regarding the role that the BECC would play in the areas to which NADBank proposes expansion:

> BECC certification should continue to be a prerequisite for NADBank financing. If not, the danger is that like most development banks, environmental assessments will be a cursory review rather than a fundamental part of the process, and that concepts of sustainability, appropriate technology and public participation and approval will not be considered.[53]

The TCPS position is consistent with a recent assertion of the U.S. General Accounting Office (GAO) on the importance of the BECC relative to NADBank: "The [BECC] reviews projects to certify that they meet established criteria for technical and financial feasibility, are environmentally sound and self-sustaining and are supported by the public. [NADBank] was established to provide financing for project certified by the [BECC]."[54] The GAO also recommended (and the TCPS endorsed) that the BECC develop a Border Infrastructure Strategic Plan to include, among other things, a set of measurable goals and "milestones" to facilitate evaluation of progress made by the BECC and NADBank in ameliorating environmental infrastructure needs at the border.[55]

Although the BECC's project review activities and potential to facilitate long-term planning in the border area are quite important, some feel that they do not go far enough to address serious environmental problems at the border. Rapid population growth is likely to exacerbate environmental problems at the border in the coming decades. Many experts and residents feel that current regional planning efforts in the border region are inadequate. And although the BECC has established sustainable development criteria, there are no minimum requirements for project compliance. As of August 2001, efforts were underway at the BECC to augment compliance with its sustainability criteria by defining minimum requirements.[56] Beyond compliance, the BECC's efforts to promote sustainable development of the U.S.-Mexico border region have been criticized for their failure to consider adequately the needs of future generations. Helen Ingram, a scholar of American public policy, water issues, and the U.S.-Mexico border region, argues that the BECC and NADBank are only encouraging the construction of a not-so-natural disaster in the desert.[57] She argues that environmental infrastructure needs should be addressed so that current generations rather than future generations shoulder the costs. Also, centers of employment should be located in areas with adequate water supplies to serve the population growth they are likely to aggravate indefinitely. According to Ingram, to do otherwise is to set up a cycle in which financially strapped communities seek growth in their tax base to pay back environmental infrastructure loans, and the growth is further stimulated by the dearth of employment opportunities outside of the desert border region. The new residents quickly outstrip the ability of existing environmental infrastructure to service their needs, thereby necessitating the arrangement of another loan. The cycle progressively weakens the desert environment's ability to "meet the needs of the present without compromising the ability of future generations to meet their own needs."[58]

Such concerns are beginning to surface in desert communities near the border region. For example, an editorial in a major Phoenix newspaper recognizes that, although there may be good reasons to slow the spread of urban sprawl in the desert, "ill-advised and unrealistic moratoriums and no-growth policies would kill the Golden Goose. The city [of Phoenix] is considering a concerted move to growth policies based on fiscal and urban reality. . . . The buzzword of today is in-fill."[59] Other desert residents remain steadfastly opposed to limits on growth. Ignoring environmental concerns, proponents of this view argue that growth control is inherently selfish: "The current Save the Desert agenda is a very popular bandwagon to jump on, but let's be honest. We have lots of desert. Yes, the infrastructure for all our new neighbors costs. So what? To all of those who are complaining, did you forget that the infrastructure was provided for you when you bought your house?"[60]

The border is also home to a coalition of fifty grassroots organizations (from the U.S., Mexican, and Pueblo nations) that work to incorporate into development planning the sort of environmental considerations ignored by such sentiments. The Rio Grande/Rio Bravo Basin Coalition, for example, aims to "help local communities restore and sustain the environment, economies, and social well-being of the Rio Grande/Rio Bravo Basin. . . . [The members of the coalition] believe that building coalitions across borders is the best way to solve international environmental problems and that our organization is a model for such cooperation."[61] In support of this purpose, the Rio Grande/Rio Bravo Basin Coalition sponsored a conference in Ciudad Juarez titled "Finding the Balance: Water and Growth in the Rio Grande/Rio Bravo Basin." The conference aimed to develop media messages, educational materials, and dialogue among formal and informal border institutions and sectors. The panels brought together a diverse range of perspectives and expertise, ranging from an ambassador with the Mexican Foreign Service to a farmer from the border region.[62] Also represented on the panels were the BECC and the International Boundary and Water Commission (IBWC), the long-standing body for dealing with U.S.-Mexican interstate water relations.[63]

The BECC conducts its own efforts to better balance economic growth and environmental concerns in the border region. Toward this end, the BECC has sought to augment the project evaluation criteria for sustainability by strengthening its voluntary "high sustainability recognition" procedures.

The goal of BECC's [Sustainable Development Work Group] program is to promote sustainable development along the border by encouraging and assisting project sponsors and communities (including those for projects already certified) to better understand and integrate economic, social and environmental sustainability early in the project planning process. This is particularly important given the limit of critical natural resources such as water on the border as well as the need for increasing the quality of life for border residents. BECC's Sustainable Development Guidance Document is available to assist communities in assessing overall project sustainability through community involvement, applying sustainable indicators, and weighting project alternatives.[64]

The BECC has also been involved in a cooperative effort with the U.S. Environmental Protection Agency (EPA), Mexico's Secretariat for the Environment, Natural Resources, and Fishing, and the U.S.-Mexico Chamber of Commerce to establish principles of sustainable development for the border region. As a result of this cooperation, these organizations have confirmed their support in an agreement titled "U.S./Mexico Business and Trade Community: The Seven Principles for Environmental Stewardship for the 21st Century." The agreement urges businesses to take the following actions:

1. "Make substantive top management commitments to . . . pollution prevention, energy efficiency, adherence to appropriate international standards, environmental leadership, and public communications";
2. "Implement innovative environmental auditing, assessment, and improvement programs";
3. "Through open and inclusive processes, develop and foster implementation of environmental management systems";
4. "Develop measures of environmental performance . . . and tie results to actions in improving environmental performance";
5. "[V]oluntarily make available to the public information on the organization's environmental performance and releases";
6. "Work with other countries operating in the same region or industry sub-sector to improve industry-wide . . . overall environmental performance"; and
7. "Promote and give support to environmental stewardship and sustainable development in the community in which the organization operates."[65]

This list has been endorsed by a number of environmental and business groups in the United States and Mexico, including the Border Trade Alliance, the National Council of Industry Chambers, the local environmental group Puebla Verde, and the U.S. Environmental Law Institute. It was also used to organize a number of workshops among private- and public-sector leaders, for the purpose of recommending policy changes to the Fox and Bush presidential administrations.[66]

Enhancing Effectiveness

The capacity to incorporate disparate views of the border helps to provide political support and stability for the new institutions. In this way, both internal and external actors can more easily find themselves predisposed to see these institutions as legitimate and helpful.[67] To maintain this support over the long term, the institutions must appear to act in a way that is consistent with dominant interpretations of the U.S.-Mexico border environment—but they must also be effective.[68] To advance the goal of improving water quality and wastewater management in the border region, NADBank and the BECC have generated an impressive level of environmental clean-up activity during their first six years of operation (1994–2000). Still, millions of dollars earmarked for this purpose have gone unused and many environmental needs remain unmet. Government agencies in both the United States and Mexico argue that these institutions have not done enough to address growing needs. Efforts to promote sustainability in the border region tend to focus on voluntary actions at the level of local government and in the business sector, with assistance and recognition available through the BECC. Recommendations for change supported by the U.S. GAO and the TCPS, among others, include finding ways to make NADBank loans more affordable, continued linking of BECC certification to NADBank loans, and development of a Border Infrastructure Strategic Plan by the BECC.

Effectiveness is particularly important in light of past experience. In the Ambos Nogales region straddling the border between Arizona and Sonora, for example, decades-long efforts by the IBWC to expand wastewater treatment facilities have not met the growing needs of these border communities.[69] The commitment of the new institutions to break from the practice witnessed in earlier efforts was strongly questioned in the case of their Mexicali II project, intended to help clean up the New River, which flows from Mexicali into the United States. The proposed site for the new wastewater treatment facility was vigorously contested by residents who, drawing on past experience, expected the facility to be a malodorous and ineffective addition to their neighborhood.[70] Under such circumstances, improvement in wastewater treatment holds significant potential to strengthen trust and cooperation in the border region. This is true as long as observable outcomes of the new institutions improve public health in a way that does not detract from resources available to poorer states elsewhere in Mexico and works to ameliorate inequalities in the border region.

Strengthening Post-Westphalian Governance?

The Westphalian approach to peace is through a balance of power between militarized states. As noted by Ken Conca in Chapter 1, stability among such entities may be enhanced through the reduction of uncertainty (related to motivations and conditions), promoting diffuse forms of reciprocity (replying in kind to another's actions), and "lengthening the shadow of the future" (expectations for productive long-term interactions). Further progress in the development of peace can be promoted through a shift in focus from state to society. Such a "post-Westphalian" approach to peace should focus on interdependence between communities, shared norms, a strong transnational civil society, transparency in government processes, and democratic accountability.[71]

A key premise of this volume is that international environmental policies should be assessed not only in terms of their ecological impact, but also in terms of their impact on peaceful cooperation among nations and across societies. To what extent have NADBank and the BECC strengthened peaceful relations in the U.S.-Mexico border region? How do their activities enhance interstate relations? How do they strengthen post-Westphalian peacemaking activities? How do they promote trust in the people and government on the "other side" of the border? Here, too, these environmental infrastructure institutions have made important contributions to strengthening ties between the governments and the societies of the United States and Mexico. Still, greater effort is needed to promote bilateral stability and post-Westphalian governance.

At the bi-national level, NADBank and the BECC are framed as contributing to improved trust and mutual responsiveness between Mexico and the United States. As a result of a meeting on border issues in September 2001, the presidents of both countries characterized U.S.-Mexico relations in the following terms:

> Both Presidents agreed that U.S.-Mexican relations have entered their most promising moment in history. Our governments are committed to seizing the opportunities before us in this new atmosphere of mutual trust. The depth, quality and candor of our dialogue is unprecedented. It reflects the democratic values we share and our commitment to move forward boldly as we deepen this authentic partnership of neighbors.[72]

Achieving these goals involved not only new institutions but new institutional cultures as well. To join the NAFTA transnational "community-region,"[73] Mexico was encouraged to accept prevailing industrialized-country ideas regarding the preferred way to facilitate equitable and effective performance: the new institutions were to follow processes consistent with institutional transparency, public participation, and sustainable development. Employment of shared norms in this bilateral context held significant potential to enhance shared understandings and mutual trust, two important elements toward the construction of a zone of peace. As Emanuel Adler notes:

> Security community-building institutions are *innovators,* in the sense of creating the evaluative, normative, and sometimes even causal frames of reference. This type of institution may also play a critical role in the *diffusion* and *institutionalisation* of values, norms, and shared understandings. Finally, by establishing norms of behavior, monitoring mechanisms, and sanctions to enforce those norms, all of which encourage, and also depend on, mutual responsiveness and trust, security community-building institutions may help shape the practices of states that make possible the emergence of security communities.[74]

Prior to the creation of NADBank and the BECC, water problems along the U.S.-Mexico border were addressed by negotiation between the U.S. and Mexican governments through the IBWC and its Mexican counterpart, the Comision Internacional de Limites y Aguas (CILA). The fact that wastewater treatment was managed through diplomatic channels was blamed for the tardy and inadequate management of transborder sewage problems.[75] During the NAFTA debate, the TCPS criticized the IBWC for failing to make its process of deliberation and decision making open to the public.[76] The BECC and NADBank were created in part to redress such concerns.

The BECC and NADBank are neither U.S. nor Mexican institutions; they are bilateral institutions. Their creation, governance, and operation were designed to require a balanced contribution and an ongoing cooperative dialogue. In contrast to the IBWC and the CILA, there is no Mexican section or U.S. section of the BECC or NADBank. Rather than splitting the organizational structure along national lines, the BECC and NADBank are split along functional lines, with NADBank financing projects certified by the BECC. Cross-border communication is also embedded into these sister institutions by their location (the BECC in Mexico and NADBank in the United States).

Communication and coordination between these sister institutions has been strained at times, but their troubles are more easily construed as interinstitutional tensions than interstate divisions. This organizational structure has reduced uncertainty between the United States and Mexico regarding what the problems entailed and how each government intended to address them. Joint commitment to these institutions has also added a new element to the set of productive collaborative activities engaging the attention of the two governments. With the addition of the BECC and NADBank to the institutions dealing with environmental concerns in the border region, the permanence of cooperation on environmental matters has reached a new level. Because the BECC and NADBank are neither here nor there— they cannot be divided along political boundaries—they contribute to the kind of firsthand intercultural experience that reduces uncertainty, promotes diffuse forms of reciprocity, and builds expectations for productive, long-term interactions.

Another innovation embedded in the BECC and NADBank is the expectation that the public should participate in policy and project decisions. This position is in marked contrast to border community perceptions of the IBWC practices prior to NAFTA. For example, in 1992, an NGO focusing on border issues made the following complaint:

There is no public participation in [IBWC] decision-making, with the exception of Environmental Impact Statements for major construction projects. Commission meetings are closed to the public. The U.S. Section of the IBWC has acknowledged that it is subject to the Freedom of Information Act. However, it treats almost all information from Mexico as confidential and it has an antiquated record-keeping system that makes it difficult for the public to get complete responses to [U.S. Freedom of Information Act] requests.[77]

The BECC cannot certify a project that is not supported by the community it is supposed to benefit. Without BECC certification, NADBank cannot fund the project. Major policy decisions of the BECC (e.g., revision of the high-sustainability criteria) and NADBank (e.g., revision of the scope of lending activities) are also subject to public scrutiny. To enable citizens to comment on individual projects and proposed policy changes, transparency in BECC and NADBank processes is necessary: the public must have ready access to relevant information before they can make a considered decision on whether to support a given project or proposal. NADBank and the BECC are funded by taxpayers in both countries, and pollution on one side of the border influences conditions on the other. As a result, concerned citizens and NGOs on either side of the border have an incentive to work with one another in their efforts to influence policy and project design at the BECC and NADBank. Through their commitment to public participation, the BECC and NADBank contribute to shared norms, transparency in government processes, interdependence between communities, an active transnational civil society, and democratic accountability.

In practice, the BECC and NADBank put considerable effort into reaching a broad and diverse sample of state and local governments, engineering consulting firms, chambers of commerce, organized interest groups, and poor, middle-income, and wealthy citizens. For example, representatives from NADBank regularly travel throughout the border region to meet with communities and local governments soliciting feedback on NADBank's policies and practices. The BECC requires that all projects demonstrate community participation and public support in order to receive BECC approval. The BECC's Community Participation Criteria state that

> applicants must submit and implement a BECC-approved Community Participation Plan that will consist of a local steering committee, meetings with local organizations, public access to project information, and at least two public meetings. . . . Following implementation of the Comprehensive Community Participation Plan, applicants must submit a report to the BECC demonstrating public support for the project.[78]

Attendance at BECC project meetings varies widely. At the first planning meeting for the Mexicali II project, seven hundred people attended.[79] A number of projects in El Paso, in contrast, generated little public interest. Some participants in one of the BECC's earliest meetings complained that the process seemed more like an assembly line than a dialogue, as each member of the public was limited to two minutes, with little commentary from the board.[80] Public participation as it is understood in the United States is a new mode of operation in Mexico. The BECC and NADBank are becoming more proficient as they gain experience and feedback on their management of past public meetings.

Another norm that the BECC and NADBank are supposed to promote in the U.S.-Mexico border region is sustainable development. Although both countries have signaled support for this concept in numerous international contexts, much work remains to be done to move this idea from promise to implementation. In a place where compliance with existing domestic environmental law is uncertain, efforts to encourage the private sector and local governments to do something even less common—"sustainable development"—would be wise to tread cautiously. At the risk of failing to tread at all, the most serious efforts to promote sustainability in the region focus on information, dialogue, symbolic commitments, and celebrated voluntarism. For example, minimal requirements for the BECC's sustainability criteria have yet to be set,[81] and the voluntary high-sustainability criteria are still far from being effectively implemented by the BECC.[82]

As frustrating as the glacial pace of progress in this area may be, this approach is consistent with Galtung's admonition that efforts to address structural violence should not be imposed upon society: "People should work out their own priorities, and the self-reliant ones will always have the courage to do so in dialogue with others."[83] To do otherwise would contradict efforts to strengthen civil society, interdependence between communities, and democratic accountability. Government innovation is constrained by the limits of shared norms and collective identities. The failure of the border communities to seek the high-sustainability recognition indicates a need for continued dialogue.

Conclusions

There is a highly unequal distribution of power in many parts of the U.S.-Mexico border region, creating a dynamic analogous to Galtung's description of structural violence.[84] There are some NAFTA-related efforts under way in the area of water and wastewater infrastructure and cooperation that have the potential to reduce these structural problems and to deepen participation, trust, and interdependence in the border region. The performance of these new institutions to date has been uneven, with some activities doing more to build diffuse reciprocity and collective identity than others. More can be done to help build peace-enhancing post-Westphalian governance mechanisms in the border region.

NADBank and the BECC have made some headway in the decades-long task of improving environmental infrastructure in the U.S.-Mexico border region. While the BECC has struggled to find resources for its activities, NADBank has been criticized for doing too little with the millions it has available to lend. At the root of the problem is the requirement that NADBank lend its money at market rates. Few border communities find such terms attractive and many find them unaffordable. So they take the EPA's grant money instead. To avoid the destabilizing effects of appearing to be unable to serve the purpose for which it was created, NADBank is trying to change its purpose: Maybe Los Angeles would like to borrow some money? Maybe a high-speed rail project would find the interest rate attractive? Maybe the interest rate could be lowered?

Before NADBank can reinvent itself, however, it has to take stock of its rhetorical roots. NADBank and the BECC were designed to balance competing views of the border environment. Among them are the view that the border poses a public health threat, the view that the border is a relatively wealthy and self-reliant region, and the view that the border is a crossroads of inequity. Any adjustment to its purpose must not only make NADBank more effective, but also make sense in terms of these dominant perspectives.

Because effectiveness and transparency are part of the cultural framework of the North American "community-region," NADBank cannot easily navigate the conflicting demands for getting the job done and satisfying popular myths as to how to do it right. Stuck in the middle, the public meetings held to solicit commentary on whether NADBank's mandate should be expanded demonstrate that the bank has thrown the dilemma out to the public. Here, then, is an opportunity for public dialogue on the kind of development bank the people of the border region would like to have available. Here is an opportunity for the type of transnational dialogue that can promote the post-Westphalian governance experiences conducive to changing the character of encounters along the U.S.-Mexico border from a zone of endemic conflict and uneasiness into a zone of peace.

Notes

1. Betty Williams, as quoted in Helena Cobban, *The Moral Architecture of World Peace: Nobel Laureates Discuss our Global Future* (Charlottesville: University Press of Virginia, 2000), 104, 106.

2. Johan Galtung, "The Basic Needs Approach," in Katrin Lederer, ed., *Human Needs* (Cambridge, Mass.: Oelgeschlager, Gunn, & Hain, 1980).

3. This banner was created by Alfonso Lorenzana and titled "Inciertas senales de una frontera" (Uneasy signs of a border).

4. This banner was created by Marco (Erre) Ramirez and titled "Rompe Cabezas/Se busca Paloma Blanca" (Puzzle/In search of the dove).

5. Interview with Marco (Erre) Ramirez., Tijuana, Mexico, June 2000.

6. For a sampling of the pressures aggravated by the maquiladoras in Baja California, see Chris Krauli, "Trouble in Boomtown," *Los Angeles Times Magazine,* May 14, 2000, 22.

7. See "In with the Tide: A Mexican Sewage Plant and Beach Pollution," *San Diego Union-Tribune,* May 12, 2000, B-8. Information on pollution of beaches near the border is available at ⟨healthebay.org⟩. Announcement of this website was deemed important enough to make it into San Diego's major newspaper: Terry Rodgers, "Beachgoers Can Surf Internet for Safety Update: Environmental Group's Site Lists Levels of Water Pollution," *San Diego Union-Tribune,* July 17, 2000, B-1.

8. U.S. Environmental Protection Agency (EPA) and the Mexican Secretariat for

Environment, Natural Resources, and Fisheries, *United States–Mexico Border Environmental Indicators* (Washington, D.C.: EPA, 1997), chap. 5, 1, available at ⟨www.epa. gov/usmexicoborder/indica97/chap5.htm⟩ on September 29, 2000.

9. Interview with Ramirez.

10. Interview with Carmen Ramirez Ruiz, Tijuana, Mexico, June 2000. For a critical analysis of the treatment of women by maquiladora employers, see Elvia R. Arriola, "Voices from the Barbed Wires of Despair: Women in the Maquiladoras, Latina Critical Legal Theory, and Gender at the U.S.-Mexico Border," *DePaul Law Review* 49 (Spring 2000): 729.

11. Galtung, "Basic Needs Approach," 64.

12. Ibid., 56.

13. Ibid., 56, 64.

14. See Robert Collier, "Mexico's Fox Urged to Adopt Green Issues; Controversial Adviser Puts Ecology on Agenda," *San Francisco Chronicle,* July 6, 2000, A1.

15. For an analysis of the governance record of the Partido Accion Nacional (National Action Party) in Baja California, see Ken Ellingwood, "Baja Offers Mexico Lessons on Opposition-Party Rule," *Los Angeles Times,* June 13, 2000, A-1, A-8.

16. See Ken Conca, "In the Name of Sustainability: Peace Studies and Environmental Discourse," *Peace and Change* 19, no. 2 (April 1994): 91–113.

17. Susan Mains, "An Anatomy of Race and Immigration Politics in California," *Journal of Social and Cultural Geography* 1, no. 2 (December 2000): 143–54.

18. Confidential interview with a leading member of a Mexico-based environmental nongovernmental organization (NGO), Ciudad Juarez, Mexico. For more information on toxic pollutants in the border environment, see the Arizona Toxics Information website: ⟨www.primenet.com/~aztoxic/⟩. See also the citizen submission to the Commission for Environmental Cooperation regarding an abandoned lead smelter in Tijuana at ⟨www.cec.org⟩.

19. Texas Center for Policy Studies (TCPS), "NAFTA and the U.S.-Mexico Border Environment: Options for Congressional Action" (Austin: TCPS, September 1992).

20. Problems continue in these rivers today. See, for example, Valerie Alvord, "Toxic River Becomes Path to USA," *USA Today,* May 11, 2000, 1A; Elvia Diaz, "Mexican Sewage a Foul Deal in Nogales," *Arizona Republic,* May 1, 2000, B-1; and Leslie Wolf Branscomb, "Sewage Woes Upset Imperial Beach," *San Diego Union-Tribune,* April 13, 2000, B-3.

21. Günther Bæchler suggests that the net result is more pollution moving from the United States to Mexico than vice versa: "Water degradation occurs mainly on the US side, whereas Mexican users' access to freshwater resources is restricted by expansion of irrigation, overuse of fertilizers, insecticides and pesticides, discharge of industrial waste, chemical pollution, and untreated sewage." Günther Bächler, *Violence through Environmental Discrimination* (Dordrecht: Kluwer, 1999), 204.

22. W. R. Hendee, "A Permanent United States–Mexico Border Environmental Health Commission," *Journal of the American Medical Association* 263, no. 24 (1990): 3319–21.

23. Copley News Service, "House OKs Funding to Fight Border Sewage; House Approves $500,000 to Help Fight Sewage at Border," *San Diego Union-Tribune,* June 27, 2000, B-5.

24. Confidential interview, Secretaria de Desarrollo Social (Secretariat of Social Development) official, Mexico City, March 1998.

25. For example, "The Sierra Tarahumara is a region of great contrasts, co-existing

cultures, and overlapping economic activities, including mining, forestry, tourism, and the drug trade. . . . Because of increased outside activity in the region, conflicts have resulted as indigenous groups seek to protect their cultural and social relationship to the land." Maria Teresa Guerrero, Cyrus Reed, and Brandon Vegter, "The Forest Industry in the Sierra Madre of Chihuahua" (Chihuahua: Comision de Solidaridad y Defensa de los Derechos Humanos, A.C.; and Austin: TCPS, April 2000), 5.

26. A respondent to a confidential survey of self-selected BECCnet subscribers and their colleagues conducted by the author during 1997–99 stated, "The most grave problem that I see for realizing sustainable development is the imbalance in develop-ment of the countries. Many poor countries surrender their resources to rich countries [who exploit them in an unsustainable manner] as the only alternative for their develop-ment. To achieve sustainable development at the global level an [amount of] equity and economic equality at the planetary level is necessary."

27. President-elect Vicente Fox, as quoted by Traci Carl, "Mexico's Fox Promotes Opening U.S. Border, Sparking Debate," Associated Press, July 17, 2000.

28. Interview with Ramirez Ruiz.

29. Gregory Alan Gross, "Border Patrol Agents to Get Water-Rescue Training," *San Diego Union-Tribune,* June 28, 2000, B-4.

30. Associated Press, "Two Rio Grande Drowners Found," *San Diego Union-Tribune,* June 11, 2000, A-23.

31. "Contra punto," Canal 11, Tijuana, June 11, 2000.

32. See Jennifer Bowles, "Lawsuit to Test Rights to Water," *Press-Enterprise* (Riverside, Calif.), June 28, 2000, A-1.

33. See Alfonso Andres Cortez Lara, "Dinamicas y conflicto por las aguas trans-fronterizas del Rio Colorado: el proyecto All-American Canal y la sociedad hidraulica del Valle de Mexicali [Dynamics and conflict regarding the trans-border waters of the Colorado River: The All-American Canal project and the hydraulic society of the Mexicali Valley]," *Frontera Norte* 11, no. 21 (January–June 1999): 33–60.

34. David Harmon, "Beset by Drought, Valley Says Mexico Is Hoarding Water," *Austin American-Statesman,* March 24, 2000, B-1.

35. For an in-depth analysis of the assumptions embedded in the NADBank and BECC institutions, see Pamela Doughman, "Discourse and Water in the U.S.-Mexico Border Region" in Joachim Blatter and Helen Ingram, eds., *Reflections on Water: New Approaches to Transboundary Conflicts and Cooperation* (Cambridge, Mass.: MIT Press, 2001).

36. Jeffrey L. Dunoff, "Institutional Misfits: The GATT, the ICJ, and Trade-Environment Disputes," *Michigan Journal of International Law* 15 (1994): 1058.

37. Diane E. Lewis, "Unions to Fight NAFTA," *Boston Globe,* December 18, 1992, 73.

38. General Agreement on Tariffs and Trade, *Dispute Settlement Panel on United States Restrictions on Imports of Tuna,* 30 I.L.M. 1594 (1991).

39. "Agreement between the Government of the United States of America and the Government of the United Mexican States Concerning the Establishment of a Border Environment Cooperation Commission and a North American Development Bank," November 1993. For additional information on the NADB and the BECC see ⟨www. nadbank.org⟩ and ⟨www.cocef.mx⟩.

40. "Border Environment Cooperation Commission Project Certification," Ciudad Juarez, BECC, June 2000; Board Resolution 2000-5, "Utilizing NADB Lending

Capacity: Implications for Mandate Expansion," approved July 11, 2000, NADB board meeting, San Antonio, Tex.

41. NADBank, "Press Release: North American Development Bank to Explore Options for Utilizing Its Lending Resources," NADBank public annual meeting, July 12, 2000. See also Cyrus Reed and Mary Kelly, "Expanding the Mandate: Should the Border Environment Cooperation Commission and North American Development Bank Go beyond Water, Wastewater, and Solid Waste Management Projects and How Do They Get There? Comments on Utilizing the Lending Capacity of the NADB" (Austin: TCPS, July 2000), 6.

42. As quoted by Steve Taylor, "Bank Set Up under NAFTA Has 'Failed Miserably,' " *Valley Morning Star* (Harlington, Tex.), August 13, 2001.

43. Reed and Kelly, "Expanding the Mandate."

44. Taylor, "Bank Set Up under NAFTA."

45. Board Resolution 2000-5.

46. NADBank, "North American Development Bank Explores Options for Utilizing Its Lending Resources: Request for Comments," press release posted to the BECCnet listserv, BECCNET8stserv.arizona.edu, July 17, 2000.

47. "Joint Statement between the United States of America and the United Mexican States" (Washington, D.C.: White House, September 6, 2001).

48. Diane Lindquist, "Bank Wants Border Area to Benefit from Development," *San Diego Union-Tribune,* June 1, 2000, C-1.

49. "Meeting" posted to BECCnet listserv by Cyrus Reed of TCPS, July 11, 2000.

50. Reed and Kelly, "Expanding the Mandate," 2, 13–16.

51. Ibid., 3, 16–17.

52. "Joint Statement."

53. Reed and Kelly, "Expanding the Mandate," 17.

54. U.S. GAO, "U.S.-Mexico Border: Despite Some Progress, Environmental Infrastructure Challenges Remain," GAO/NSIAD-00-26 (Washington, D.C.: GAO, March 2000), 3.

55. U.S. GAO, "U.S.-Mexico Border," 29, as cited in Reed and Kelly, "Expanding the Mandate," 6.

56. BECC, "Proposal for the Establishment of Minimum Requirements for Project Compliance with BECC's Sustainable Development Criteria," August 2001, available at ⟨www.cocef.org⟩ on September 2, 2001.

57. Helen Ingram, "Water and the Globalizing Economy: The Coming Crisis," in Uriel Rosenthal, Arjen Boin, and Louise K. Comfort, eds., *Managing Crises: Threats, Dilemmas, Opportunities* (Springfield, Ill.: Charles C. Thomas, 2001).

58. Ingram, "Water and the Globalizing Economy." See also World Commission on Environment and Development, *Our Common Future* (New York: Oxford University Press, 1987), 43.

59. "Growth in the Valley: Is Sprawl Inevitable?" *Arizona Republic,* October 2, 1994, C-4.

60. Bob Hepker, "Growth Foes Do Not Want to Share Desert," *Arizona Republic,* July 9, 1999, 2.

61. Rio Grande/Rio Bravo Basin Coalition, "Welcome/Bienvenidos" web page: ⟨www.rioweb.org⟩ on September 30, 2000.

62. Rio Grande/Rio Bravo Basin Coalition, "Uniting the Basin 2000" web page: ⟨www.rioweb.org/UnitingtheBasin2000.html⟩ on September 30, 2000.

63. The IBWC as currently constituted was created in 1944, replacing the International Boundary Commission created in 1889.

64. Sustainable Development Work Group, "Draft: New Sustainable Development Work Plan for Public Review and Comments," (Ciudad Juarez: BECC, June 20, 2000), 3, available at ⟨www.cocef.org/⟩ on September 30, 2000.

65. Lawrence I. Sperling, Charles Cervantes, Carlos Gonzalez Guzman, and Ricardo Castañón, "The Seven Principles of Environmental Stewardship for the 21st Century," June 4, 1999, available at ⟨www.usmcoc.org/sevenprinciples.html⟩ on December 16, 2001. The authors are with the EPA, Mexico City; the U.S.-Mexico Chamber of Commerce; the Office of the Mexican Federal Attorney for Environmental Protection; and the BECC, respectively.

66. Sperling, Cervantes, Gonzalez Guzman, and Castañón, "The Seven Principles."

67. "Incorporating externally legitimated formal structures increases the commitment of internal participants and external constituents. And the use of external assessment criteria—that is, moving toward the status in society of a sub-unit rather than an independent system—can enable an organization to remain successful by social definition, buffering it from failure." John W. Meyer and Brian Rowan, "Institutionalized Organizations: Formal Structure as Myth and Ceremony," in Walter W. Powell and Paul J. DiMaggio, eds., *The New Institutionalism in Organizational Analysis* (Chicago: University of Chicago Press, 1991), 49.

68. "The more an organization's structure is derived from institutionalized myths, the more it maintains elaborate displays of confidence, satisfaction, and good faith, internally and externally. . . . Participants not only commit themselves to supporting an organization's ceremonial facade but also commit themselves to making things work out backstage. The committed participants engage in informal coordination that, although often formally inappropriate, keeps technical activities running smoothly and avoids public embarrassments. In this sense the confidence and good faith generated by ceremonial action is in no way fraudulent. It may even be the most reasonable way to get participants to make their best efforts in situations made problematic by institutionalized myths at odds with immediate technical demands." Ibid., 59.

69. Helen Ingram and David R. White, "International Boundary and Water Commission: An Institutional Mismatch for Resolving Transboundary Water Problems," *Natural Resources Journal* 33, no. 1 (Winter 1993): 153–75.

70. *BECC Step II Project Format for the Sanitation Program of the City of Mexicali* (Mexicali: State Public Services Commission, October 1997).

71. See Ken Conca, "Environmental Cooperation and International Peace," in Paul F. Diehl and Nils Petter Gleditsch, eds., *Environmental Conflict* (Boulder: Westview, 2000).

72. "Joint Statement."

73. "Community-regions are regional systems of meanings . . . made up of people whose common identities and interests are constituted by shared understandings and normative principles other than territorial sovereignty and (a) who actively communicate and interact across state borders, (b) who are actively involved in the political life of an (international and transnational) region and engaged in the pursuit of regional purposes, and (c) who, as citizens of states, impel the constituent states of the community-region to act as agents of regional good, on the basis of regional systems of governance." Emanuel Adler, "Imagined (Security) Communities: Cognitive Regions in In-

ternational Relations," *Millennium: Journal of International Studies* 26, no. 2 (Summer 1997): 249–77, at 253.

74. Ibid., 268.

75. Ingram and White, "International Boundary and Water Commission."

76. "NAFTA and the U.S./Mexico Border Environment: Options for Congressional Action" (Austin: TCPS, 1992), 3-1.

77. Ibid.

78. "BECC Project Certification Criteria," (Ciudad Juarez: BECC, November 9, 1996).

79. *BECC Step II Project Format,* 46.

80. Confidential interview with NGO activist, Washington, D.C., summer 1997.

81. BECC, "Proposal for the Establishment of Minimum Requirements."

82. See Sustainable Development Work Group, "Draft: New Sustainable Development Work Plan."

83. Galtung, "Basic Needs Approach," 71.

84. Ibid.

8

The Problems and Possibilities of Environmental Peacemaking

Ken Conca and Geoffrey D. Dabelko

We began this volume by suggesting that emphasis on pathways from environmental degradation to violent conflict may have crowded out inquiry and action on linkages between environmental cooperation and the enhancement of peace. We argued that there may be useful catalytic roles for environmental cooperation in peacemaking, which we define as movement along a continuum ranging from the absence of violent conflict to the unimaginability of violent conflict. We speculated further that there may be two broad pathways for moving along this continuum. One involves using environmental initiatives to generate an improved climate of intergovernmental relations; the other is rooted in peace-enhancing cross-border societal linkages that may result from environmental cooperation and efforts to promote it. If these connections can be made, then environmental cooperation may have an important role to play beyond its traditional purposes of stemming environmental degradation, enhancing human welfare, or cutting off the specific ecological pathways to violence.

What can we conclude about the peace-catalyzing possibilities of environmental cooperation from the cases examined in this volume? We begin with the usual caution: a handful of cases is a poor basis for generalization. Beyond this generic limit, we must be careful about drawing firm conclusions for a simple reason: meaningful, robust, institutionalized environmental cooperation is at best incipient in all of the regions studied. The institutional arrangements at the heart of most of our cases—be they the Caspian Environmental Program (CEP), the Aral Sea accords, the side agreements to the North American Free Trade Agreement (NAFTA), the Southern African Development Community (SADC) protocol on watercourses, or most of South Asia's bilateral water accords—have all emerged within the past decade. Even in the Baltic, the region with the longest-standing and best developed institutional infrastructure among the cases covered here, environmental cooperation has only just begun to integrate the eastern Baltic littoral states and has barely considered the question of effectively engaging post-Soviet Russia.

The case studies gathered in this volume cannot be interpreted, therefore, as formal tests of environmental peacemaking propositions. Simply put, neither governments in these regions nor influential outsiders have put our propositions to the test in more than an incipient, halting way. The World Bank played a crucial catalytic role in brokering agreement on the Aral Sea crisis, but has largely failed to build on that opening to create more robust forms of regional environmental governance. Water cooperation accords have proliferated in South Asia, but most constitute water-sharing deals that lack any meaningful characteristics of sustainable, participatory watershed governance. The NAFTA side agreement created an innovative mechanism for funding community projects, but the resulting institutions have lent only a fraction of the available funds in their first few years of operation. We are left, therefore, to ask whether environmental peacemaking *could* be a reality in these or other regions, given the partial evidence provided by these incipient social processes.

With these general observations in mind, we turn to the questions posed in Chapter 1, dealing with the specific pathways that might link environmental cooperation and regional peace. We asked first whether institutionalized environmental cooperation could forestall the catalytic or triggering processes that figure prominently in eco-violence scenarios. Here we are wary of drawing firm conclusions. Our cases were not selected to be a strong test of propositions about ecologically induced violence or its antidotes; rather, our concern has been with the catalytic effects of environmental cooperation on regional peace and conflict more generally. Nor have the cases dealt in detail with the highly localized subnational processes that are at the heart of most eco-violence.

Still, we do note that at least some of our case-study authors see environmental cooperation as having played a role in forestalling eco-violence of one form or another. Erika Weinthal in Chapter 4 points out that localized violence was on the increase in Central Asia in the aftermath of the Soviet collapse, fueled by a combustible mix of ethnicity, water scarcity, and newly drawn lines of state authority. She concludes that water-sharing cooperation around the Aral Sea and its feeder rivers helped to defuse such tensions. Ashok Swain, writing about South Asia in Chapter 3, cautions against being too quick to dismiss the possibility of water as a trigger for interstate conflict in the absence of water-sharing agreements. He sees substantial benefits to regional stability from the various bilateral water accords that have been worked out between states in the region. On the other hand, these accords have done little or nothing to address the intra-societal tensions seen by most scholars as the most likely pathway to eco-violence. The case of South Asia provides us with an important caution: Whether the concern is ecologically induced violence or regional insecurity more generally, the mere existence of cooperation is less important than the content, scope, and orientation of that cooperation.

Turning from the narrower question of eco-violence to our primary concern—the problem of regional insecurity in the absence of robustly peaceful relations—we arrive at the central question of environmental peacemaking: Are there side effects from such cooperation that can create positive synergies for peace? To answer this question we look first at the pattern of interstate dynamics revealed in the cases. We then turn to the question of trans-societal linkages and interactions outside the formal domain of interstate relations.

Interstate Dynamics

As suggested in Chapter 1, environmental problems are often marked by complex uncertainties, extended time horizons, highly nonlinear patterns of change, and overlapping ecosystemic interdependencies. Do these conditions have the effect of altering what cooperation theorists would call the "contractual environment" of interstate bargaining? If so, do they change the structure of interstate bargaining in ways conducive to peace? We posited that they might do so through processes of trust building, the creation of consensual knowledge, the establishment of longer time horizons, the identification of mutual gains, or the fostering of a growing habit of cooperation. Again, the cases presented here do not constitute rigorous tests of specific hypotheses about the ways that environmental cooperation might change the trajectory of regional interstate relations. Nevertheless, they contain several hints, suggestions, and inferences that provide important insights.

First and foremost, we note that environmental cooperation is high politics, in the sense of engaging the sustained attention of state actors at the highest level. In the Caspian region, the initial catalyst for cooperation was a strategic Iranian initiative. Perceptions that Russia uses environmental concerns as a tool to suppress Caspian states' autonomy and development continue to dog cooperative environmental efforts. In South Asia, India has used subregional environmental initiatives to isolate Pakistan and to repair relations with Bangladesh; smaller states resist some Indian forays, fearing Indian hegemony in the absence of Pakistani balancing. The early stages of Baltic Sea regional cooperation were heavily influenced by the search for cooperative political space in the context of the Cold War. In Central Asia, state elites have viewed regional resource-based cooperation as a critical way of demonstrating regional stability and their status as sovereign states. In North America, environmental cooperation became a necessary condition for the economic integration sought by political and economic elites on both sides of the U.S.-Mexico border. In southern Africa, ecotourism and "sustainable development" have emerged as critical elements of economic growth strategies and a core theme of SADC.

In other words, across all of these regions, it makes little sense to view environmental initiatives as low-stakes, "functional" cooperation akin to coordinating postal service, air traffic control, or the navigation of shared waterways. If environmental cooperation can lead to a more robust peace it is not because mundane, low-stakes environmental cooperation creates a slippery slope toward cooperation on the real issues of contention. A slippery slope may indeed exist, but we should look for it at the level of high politics rather than bureaucratic administration. The political and economic stakes in environmental cooperation are much higher—and are clearly understood to be so by all actors involved.

It is no accident that several of the cases in this volume that play out as high politics focus in some fashion on water. Water quality and quantity hold a preeminent position in any hierarchy of environmental concerns. Water is high politics because it is rightly perceived as an absolute necessity in the very near term. If other environmental concerns are understood as longer-term threats, luxury goods, or issues unrelated to economic development and people's livelihoods, they may not have the same energizing effect. This suggests the importance of finding and identifying these high-politics connections as part of environmental initiatives in order to maximize their peacemaking potential, whether the issue in question is water, forests, air quality, or toxic contamination.

The status of environmental cooperation as high politics is no doubt a double-edged sword. On the one hand, it means that robust, meaningful environmental cooperation may be harder to attain. Environmental cooperation may pose threats to well-entrenched interests; at the very least, complex issue linkages must be managed. One the other hand, these conditions suggest that cooperation—once established—may generate broader leverage for change than would some narrower, lower-stakes domain of functional cooperation.

This observation is important because, without minimizing the difficulties of establishing robust environmental cooperation, we note a second common pattern across the cases: environmental cooperation can be a useful entry point in the context of interstate political conflict. In South Asia, India and Pakistan found a way to cooperate over the Indus River and to sustain that cooperation through three shooting wars. In the Caspian Sea region, political space has emerged for the CEP despite deep geopolitical tensions and raging conflicts over resource property rights. In the Baltic Sea region, environmental cooperation emerged from the inhospitable strategic and political climate of the Cold War to become the most important strand of regional cooperation—indeed, arguably the only significant strand during that era.

If environmental cooperation is indeed high politics, then we cannot dismiss the existence of environmental cooperation in otherwise tense or conflictual settings as insignificant. Instead, this finding suggests that environmental links around shared watercourses, regional seas and airsheds, border ecology, and transboundary pollution may be sufficiently immediate and contentious to draw states and societies into meaningful regional cooperation—but not so immediate and contentious as to create impossible barriers to collective action.

Stacy VanDeveer's analysis of the Baltic Sea case in Chapter 2 suggests that it was possible to instill at least the beginnings of a broader habit of cooperation through environmental initiatives. Significantly, this cooperation emerged not in today's relatively conducive climate—marked by the pull of Europe and the end of the Cold War—but in the inhospitable political climate of East-West tensions. Today, regular ministerial conferences have become a well-established ritual, states face strong pressures to ratchet upward their cooperative entanglements, and the whole process has spread beyond the environment ministries to engage a wide array of state entities.

Douglas Blum's interpretation of the Caspian Sea case in Chapter 6 is guardedly optimistic with regard to a broadly similar pathway to improved interstate dynamics. He sees the CEP as lengthening the time horizon, drawing states into joint projects promising shared future gains, and helping to instill a habit of regional cooperation. One striking aspect of the Caspian case is the move toward cooperation in the absence of a framework environmental treaty among the participant states. This cooperation flies in the face of much conventional wisdom about the utility of highly general, nonspecific, even vague framework environmental accords that can then be made more precise and binding over time. Yet, given the great wariness of states in the region about questions of legal ownership and the potential legal ramifications of an environmental treaty for resource extraction, it seems unlikely that the formal-legal approach would generate the same cooperative synergies Blum sees emerging around the CEP.

Weinthal's Chapter 4 interprets the Aral Sea case as a somewhat similar story, but with an important twist. What drew states to the bargaining table was initially the reality of their shared interdependence and the need for some predictability during the tumult of post-Soviet transition. What kept them at the bargaining table, however, was not simply functional interdependencies, but rather their ability to use cooperation around the Aral and its feeder rivers in the effort to establish themselves as autonomous, sovereign, competent states. Essentially, her argument is that Aral cooperation stabilized regional relations because outside actors had made stable cooperation a prerequisite for providing crucial financial and political-symbolic resources to the newly independent states of the region. Although the case specifics may turn on the unique event of Soviet collapse, the more general argument parallels the "new sovereignty" thesis of Abram Chayes and Antonia Handler Chayes: that entering into cooperative, problem-solving accords is the defining essence of contemporary statehood in the international context.[1]

In many ways, the South Asia and southern Africa cases present tougher challenges to this sort of state-centered cooperative stabilization. To be sure, there are hopeful aspects to each case. In South Asia, the recognition of functional interdependencies has resulted in a proliferation of watercourse agreements despite enduring political tensions. In southern Africa, the attempt to build what Larry Swatuk in Chapter 5 terms "common resource regimes" has emerged as a leading instrument for the construction of a growing habit of cooperation among the SADC states. Each case hints at the possibility of new cognitive and political dynamics at work as states calculate their interests. Yet these cases also return us to the critical importance of the form and content of cooperation. In South Asia, water-sharing agreements have largely avoided broader questions of sustainable watershed management—to the point of creating fictional water in the Ganges, so that each state may claim to have gotten its perceived fair share. Swain in Chapter 3 sees very positive developments from river-based cooperation between India and Nepal around the Mahakali River, yet the social and environmental consequences of the agreement's developmentalist agenda continue to spawn bitter social contestation. In southern Africa, as Swatuk suggests, there may be a fine line between interstate "environmental" cooperation and shared resource plunder. Such initiatives may instill a habit of cooperation and identify mutual gains (at least for some segments of society). But they do little to shift the frame away from a calculus of short-term advantage for individual states. Moreover, they rest on a foundation of social control rather than societal participation, suggesting tensions with the trans-societal aspects of environmental peacemaking.

Trans-societal Dynamics

Chapter 1 underscored the importance of looking beyond the state to the trans-societal dimension of regional relations. There we posed several questions: whether environmental cooperation can help to deepen and broaden trans-societal linkages, whether it can stimulate the emergence or strengthening of regionally grounded identity constructs, and whether it might help transform institutions grounded in the zero-sum logic and opaque practices of security-minded states.

With regard to trans-societal linkages, the cases reveal a mixed picture. On the one hand, it is clear that most of the regions we have studied lack a robust, well-established set of transnational civil-societal linkages. The idea that such ties might exert pressure "from both ends" for regional cooperation remains an appealing one, but testing it in these cases would require substantially deeper and more continuous interaction than seen in most of these regions. In particular, across several of the cases, a relatively weak domestic foundation for civil society in some or all of the countries involved represents a limiting factor for the emergence of more robust cross-national ties. The ramifications of this shortcoming for some of the more optimistic notions about a transcendent global civil society are sobering.

Nevertheless, we also see at least the beginnings of a thin transnational community in several cases, rooted in scientific networks, some of the better institutionalized local citizens' groups, and extraregional actors including both intergovernmental and nongovernmental organizations

(NGOs). Blum's assessment of the CEP suggests that institutionalized co-operation can play a role in stimulating the growth of such linkages. The challenge would seem to be to tie these communities more effectively to existing or emerging domestic social networks, rather than expecting them to serve as a substitute for a robust civil society.

A second set of trans-societal questions dealt with the emergence or strengthening of regionally grounded identity constructs. Both the Baltic and Caspian cases suggest the possibility that organizing environmental governance around a conception of a shared regional ecosystem could foster associations with a regional identity construct. The authors of both case studies point to the importance in this process of disseminating the idea of a shared ecosystem with unique characteristics. Both cases have managed, for the most part, to avoid framing the issues in terms of clear winners and clear losers linked in a stark upstream-downstream relationship. The problems put at the center of environmental cooperation have facilitated a region-wide construction of the task.

Several other cases suggest cautionary notes, however. Pamela Dough-man's Chapter 7 analysis of the U.S.-Mexico border region underscores the fact that existing social differences may foster sharply different interpretations of the same ecosystemic reality (as in her discussion of the multiple frames available for interpreting border life and border environmental problems). It remains unclear whether the dominant effect of these contrasting interpretations is to provide a basis for dialogue or to undermine shared conceptions of regional "citizenship." An important consideration suggested in Doughman's analysis of the U.S.-Mexico border case is whether the institutional form taken by environmental cooperation can accommodate these contrasting (if not always conflicting) interpretations. Similar questions arise around the Baltic Sea. Clearly, the pull of "Europe" has much to do with convergence on environmental norms, policies, and procedures. Yet whether and how a regionally grounded, ecosystemically reinforced Baltic identity construct fits together with a broader conception of Europe remains unclear.

The Aral Sea case suggests a different caution about the prospects for a regional identity community grounded in environmental considerations. Weinthal points to ongoing processes, including choices made by the international community, that reinforce the Soviet-era notion of the Uzbeks as "water people" and the Kazakhs as "oil people." This reinforcement facilitates post-Soviet nation-building around the continuing problem of extractive, resource-based, monocultural economies. This pattern should caution us against easy assumptions that resource-based or ecosystemic identities will automatically come down on the side of eco-holistic sustainability and, by extension, peaceful regional relations.

Moreover, as Swatuk points out in the southern African case, there may not be that great a distance between conceptions of community grounded in regional ecological identity and the traditional machinations of state power. One hopeful trend he identifies on the southern African scene is the regional networking among environmentalists, indigenous activists, and other civil-society groups in the Okavango delta. The government of Botswana often closely mirrors these positions. Although he does suggest that using environmental arguments to support their agenda may be causing Batswana policymakers to "reconsider natural resource use patterns, and to favor sustainable and more equitable uses", he sees their behavior as informed primarily by a traditional paradigm of state power through resource capture.

A third trans-societal question involved the transformation of state institutions away from a zero-sum logic of national security and toward norms of accountability, participation, and cooperation. Here we see two very different tendencies. On the one hand, the cases provide several examples of states forced to or choosing to embrace what Karen Litfin has termed "sovereignty bargains," in which aspects of autonomous control are ceded in favor of the benefits of functional cooperation.[2] These bargains in turn create pressure on state institutions to engage with and respond to a wider array of stakeholders in a much more transparent, participatory fashion. But there are also examples of what Blum describes starkly in the Caspian case as a "lingering mentality of realpolitik, paranoia, and related competitive efforts to control information." Sometimes these countervailing tendencies can be seen in the same case—as in the Baltic region, where postcommunist governments have become far more transparent in the dissemination of pollution information than in the nuclear sector.

One obvious cautionary note in this process is the question of who may be transforming whom. As Blum suggests in the context of the CEP, "NGOs may gain leverage, standing, and access to information through their inclusion in cooperative arrangements, but may also be co-opted by the state in the latter's quest for control. . . . [d]ifferent constituencies envision markedly divergent roles for NGOs." Similarly, Weinthal points not only to the enduring control of state institutions over negotiations around Aral Sea cooperation, but also to the opportunities that environmental cooperation has created "to co-opt the language of the eco-nationalist movements that were mounting a real challenge to the authority and legitimacy of the new governments."

Linking the Two Pathways

These complex state-society dynamics raise larger questions about the relationship between the two very different pathways to regional peace via environmental cooperation that we envisioned in Chapter 1. In the Baltic case, VanDeveer stresses the interplay of trans-societal and intergovernmental channels. He cautions against trying to understand effects in one channel in isolation from the other: "In a region with so many avenues for environmental cooperation at multiple levels of governance and among many state and nonstate actors, participants may be responding to changes in both the strategic climate and societal transformation simultaneously." Can we see a pattern in the less robust institutional context of the other cases, which have yet to articulate these multiple channels and levels?

The clear implication in both the Caspian and the Aral cases is that the path of improved intergovernmental dynamics can be a foundation for stronger trans-societal linkages. This finding is not surprising in light of the simultaneous challenges of building more capable state institutions and fostering a more robust civil society. Tensions are not entirely absent, of course: there are clear limits to the tolerance of authoritarian regimes in both regions for independent societal organizing. Yet the authors of both case studies see the stabilization of regional interstate relations as a critical foundation for whatever degree of transnationalism may ultimately flourish. Weinthal sees the "state-making" and state-legitimating aspects of environmental cooperation as a necessary precursor to a more transnationalist, post-sovereign future for the Aral Sea region; Blum suggests that it may be necessary for international NGOs to "stand in" for a still-developing civil society.

In contrast, the South Asian and southern African cases caution that an improved contractual environment for intergovernmental bargaining cannot be expected to lead automatically or quickly to more robust and peace-enhancing trans-societal relations. Swain finds little evidence of a trans-societal stimulus from the intergovernmental water accords that have proliferated in South Asia. One reason for this may be that they are largely water-sharing arrangements that fail to grapple with the broader question of sustainable watershed management. The southern African case offers an even starker interpretation of the interplay between the trans-societal and intergovernmental forms of cooperation. Swatuk concludes that regional ecological interdependencies have in fact led to what he calls "common resource regimes," which contribute to a more generally improved bargaining climate in the region. But he takes little comfort in that fact, underscoring the character of those regimes as mechanisms for joint resource capture, state power, and elite privilege rather than the peace-enhancing effects that might come from working to sustain nature and secure livelihoods.

Finally, the cases as a group support VanDeveer's caution against over-conceptualizing the distinctness of these two paths. On the one hand, the improved contractual environment stressed in the discussion of inter-governmental dynamics may also be an important element in cross-border dealings among nonstate actors as well; they, too, can suffer from mistrust, short time horizons, a lack of reciprocity, and the absence of a habit of cooperation. Similarly, the identity transformations and interdependencies emphasized in conceptualizing the trans-societal pathway to peace may have a critical role to play within the state. Studies of the role of transnational advocacy in changing state policy underscore the importance of sympathetic "insiders" within the state as key nodes in transnational networks.[3] These findings suggest that "transbureaucratic" linkages may be just as important as "trans-societal" ones.

Conclusion: Caution and Opportunity

Returning to the observation with which we began this chapter, we cannot conclude that environmental cooperation causes peace. We remain convinced, however, that certain forms of environmental cooperation could be extremely useful tools in the hands of peacemakers. We should not expect that all forms of environmentally based cooperation will have peacemaking effects, any more than we should expect all forms of environmental cooperation to afford the same degree of environmental protection.[4] The key is not environmental cooperation per se, but rather specific forms of cooperation that are designed to build a habit of cooperation, transform interstate bargaining dynamics, and deepen peaceful trans-societal linkages conducive to peaceful cooperation.

We also caution against extrapolating these observations about peacemaking dynamics from the regional or transnational level of analysis to the subnational or intrastate scale. Can the environment play a confidence-building role between conflicting factions, social groups, or ethno-national communities within a given state? The interest in peacemaking at the subnational level is twofold. First, most violent conflicts at the beginning of the twenty-first century are within states and not between them. Second, existing research suggests that when environmental stress makes a contribution to violent conflict, it is most likely to be at the subnational level. Both regional and substate levels of analysis are important to crafting peace. Our emphasis on the former does offer some important hints about the latter: that interstate accords in South Asia may be inadequate to deal with intrastate insecurities, for example, or that interstate cooperation around the Aral Sea may have played a role in defusing potential intrastate violence. But the substate level of analysis for environmental peacemaking clearly represents an area for future empirical and applied research in its own right.

A third caution, and a thesis ripe for wider investigation, involves the possibility of unsustainable environmental cooperation. The problem is seen most starkly in Swatuk's chapter on southern Africa, but it is also suggested in the context of water accords in South and Central Asia. If sustainability remains an important part of the foundation for a lasting peace, then environmental cooperation that facilitates more aggressive resource exploitation or continued patterns of unsustainable use may be problematic. This possibility offers a cautionary advisory to the policymaker or practitioner attempting to draw policy design conclusions from this scholarship.

Despite these cautions, we view the growing interest in proactive environmental peacemaking initiatives—seen within the European Union, among several national governments, and on the part of many NGOs—as a hopeful trend. Although the linkage between environmental stress and violent conflict has garnered considerable policy attention over the past decade, practitioners remain frustrated by the inability to translate its lessons into specific conflict-prevention measures. Most policy initiatives are fragmented and are commonly driven by single institutions, based on their unique mandate, constituencies, institutional interests, and reach.

Also, the environment-security debate has focused heavily on traditional Northern security concerns emanating from environmental stress and population growth. This focus has not easily translated into a positive, practical policy framework for cooperation and environmental peacemaking. It has largely failed to engage a broad community of stakeholders, particularly in the South. Fashioning a proactive agenda starts with developing a constructive dialogue among Northern and Southern practitioners and scholars as a means to provide substantial—and, in the eyes of the South, legitimate—input into key international forums.[5]

An environmental peacemaking agenda also offers an opportunity to overcome the commonly held critique of environmental security as currently constituted: that the emphasis on security brings with it an undue and counterproductive emphasis on conflictual methods, frameworks, and institutions. Whether enunciated by Northern national governments or by intergovernmental organizations, the specter of Northern security concerns appears to many in the South as a threat to sovereignty and a barrier to cooperative, equitable means of coping with environmental challenges.

In our judgment, the 2002 World Summit on Sustainable Development in Johannesburg, South Africa—commonly referred to as "Rio + 10"—marked the end of an era when governments could either ignore the connections among ecology, peace, and violence or filter them through an unreconstructed understanding of conflict and security. That ending presents a moment of opportunity for a strategy of more explicit environmental peacemaking efforts. By most measures, momentum is flagging in international environmental governance. Perhaps environmental peacemaking offers the chance to re-energize the process with an ambitious, inclusive set of ideas about international policymaking, combining environmental, developmental, and peace-related concerns.

Notes

1. Abram Chayes and Antonia Handler Chayes, *The New Sovereignty: Compliance with International Regulatory Agreements* (Cambridge, Mass.: Harvard University Press, 1995).

2. Karen Litfin, "Sovereignty in World Ecopolitics," *Mershon International Studies Review* 41, no. 2 (November 1997):167–204. See also Ken Conca, "Rethinking the Ecology-Sovereignty Debate," *Millennium: Journal of International Studies* 23, no. 3 (January 1994): 701–11.

3. Margaret Keck and Kathryn Sikkink, *Activists beyond Borders: Advocacy Networks and International Politics* (Ithaca: Cornell University Press, 1998).

4. On the effectiveness of international environmental accords, see Thomas Bernauer, "The Effect of International Environmental Institutions: How We Might Learn More," *International Organization* 49, no. 2 (Spring 1995): 351–77; and Oran R. Young and Marc A. Levy with Gail Osherenko, "The Effectiveness of International Environmental Regimes," in Oran R. Young, ed., *The Effectiveness of International Environ-*

mental Regimes: Causal Connections and Behavioral Mechanisms (Cambridge, Mass.: MIT Press, 1999).

5. An interesting example is Youba Sokona et al., "A Southern Dialogue: Articulating Visions of Sustainable Development," Aviso information bulletin on global environmental change and human security no. 5 (September 1999).

Index